AUTOCAD® 2008 ESSENTIALS

LICENSE, DISCLAIMER OF LIABILITY, AND LIMITED WARRANTY

The CD-ROM that accompanies this book may only be used on a single PC. This license does not permit its use on the Internet or on a network (of any kind). By purchasing or using this book/CD-ROM package (the "Work"), you agree that this license grants permission to use the products contained herein, but does not give you the right of ownership to any of the textual content in the book or ownership to any of the information or products contained on the CD-ROM. Use of third party software contained herein is limited to and subject to licensing terms for the respective products, and permission must be obtained from the publisher or the owner of the software in order to reproduce or network any portion of the textual material or software (in any media) that is contained in the Work.

INFINITY SCIENCE PRESS LLC ("ISP" or "the Publisher") and anyone involved in the creation, writing or production of the accompanying algorithms, code, or computer programs ("the software") or any of the third party software contained on the CD-ROM or any of the textual material in the book, cannot and do not warrant the performance or results that might be obtained by using the software or contents of the book. The authors, developers, and the publisher have used their best efforts to insure the accuracy and functionality of the textual material and programs contained in this package; we, however, make no warranty of any kind, express or implied, regarding the performance of these contents or programs. The Work is sold "as is" without warranty (except for defective materials used in manufacturing the disc or due to faulty workmanship);

The authors, developers, and the publisher of any third party software, and anyone involved in the composition, production, and manufacturing of this work will not be liable for damages of any kind arising out of the use of (or the inability to use) the algorithms, source code, computer programs, or textual material contained in this publication. This includes, but is not limited to, loss of revenue or profit, or other incidental, physical, or consequential damages arising out of the use of this Work.

The sole remedy in the event of a claim of any kind is expressly limited to replacement of the book and/or the CD-ROM, and only at the discretion of the Publisher.

The use of "implied warranty" and certain "exclusions" vary from state to state, and might not apply to the purchaser of this product.

AUTOCAD® 2008 ESSENTIALS

By
Munir M. Hamad
Autodesk Approved Instructor

INFINITY SCIENCE PRESS LLC
Hingham, Massachusetts
New Delhi

Copyright © 2008 by INFINITY SCIENCE PRESS LLC
All rights reserved.

This publication, portions of it, or any accompanying software may not be reproduced in any way, stored in a retrieval system of any type, or transmitted by any means or media, electronic or mechanical, including, but not limited to, photocopy, recording, Internet postings or scanning, without prior permission in writing from the publisher.

INFINITY SCIENCE PRESS LLC
11 Leavitt Street
Hingham, MA 02043
Tel. 877-266-5796 (toll free)
Fax 781-740-1677
info@infinitysciencepress.com
www.infinitysciencepress.com

This book is printed on acid-free paper.

Munir M. Hamad. *AutoCAD® 2008 Essentials.*
ISBN: 978-1-934015-06-3

The publisher recognizes and respects all marks used by companies, manufacturers, and developers as a means to distinguish their products. All brand names and product names mentioned in this book are trademarks or service marks of their respective companies. Any omission or misuse (of any kind) of service marks or trademarks, etc. is not an attempt to infringe on the property of others.

Library of Congress Cataloging-in-Publication Data

Hamad, Munir M.
  AutoCAD 2008 essentials / Munir M. Hamad.
    p. cm.
  Includes bibliographical references and index.
  ISBN-13: 978-1-934015-06-3 (hardcover with cd-rom : alk. paper)
  1. Computer graphics. 2. AutoCAD. I. Title.
  T385.H3293 2007
  620'.00420285536—dc22

2007036967

Printed in the United States of America
8 9 10 5 4 3 2 1

Our titles are available for adoption, license or bulk purchase by institutions, corporations, etc. For additional information, please contact the Customer Service Dept. at 877-266-5796 (toll free).

Requests for replacement of a defective CD-ROM must be accompanied by the original disc, your mailing address, telephone number, date of purchase and purchase price. Please state the nature of the problem, and send the information to INFINITY SCIENCE PRESS, 11 Leavitt Street, Hingham, MA 02043.

The sole obligation of INFINITY SCIENCE PRESS to the purchaser is to replace the disc, based on defective materials or faulty workmanship, but not based on the operation or functionality of the product.

PURPOSE & OBJECTIVES

This book is for novice users of AutoCAD 2008. It covers the beginner and intermediate levels. The text demonstrates in a very simple, step-by-step procedure of creating an engineering drawing, modifying it, annotating it, dimensioning it, and finally printing it.

At the completion of this book, the reader will be able to:

- Understand what AutoCAD is and how to deal with its basic operations, including its filing system
- How to draw the different objects in a fast and precise manner
- How to set up drawings
- How to construct drawings in simple steps
- How to modify any object in a drawing
- How to create, insert, and edit blocks
- How to hatch using different hatch patterns and methods
- How to create text and tables
- How to insert and edit dimensions
- How to prepare and plot a drawing

CONTENTS

Chapter 1. Introduction to AutoCAD 2008   1

    What is AutoCAD?   1
    How to Start AutoCAD 2008   1
    Points in AutoCAD   2
    AutoCAD Default Settings   3
    Things You Have to Know about AutoCAD   4
    How to Issue Commands in AutoCAD   4
    DYN and Typing AutoCAD Commands   4
    Drawing Limits   5
    AutoCAD Units and AutoCAD Spaces   6
    Right-clicking   6
    Viewing Commands   8
    Create a New File   9
    Open an Existing File (Single File)   10
    Open an Existing File (Multiple Files)   11
    Save and Save As Files   13
    Exiting AutoCAD   14
    Notes   15
    Exercise 1   16
    Chapter Review   16
    Chapter Review Answers   17

Chapter 2. Drafting Using AutoCAD 2008   19

    Introduction   19
    Line command   19
    Dynamic Drafting   20
    Exercise 2   21
    Precision Method 1: Snap and Grid   22
    Exercise 3   23
    Precision Method 2: Direct Distance Entry and Ortho   24
    Exercise 4   25
    Arc Command   26
    Exercise 5   28
    Circle Command   29

Exercise 6	31
Precision Method 3: Object Snap (OSNAP)	32
Exercise 7	35
Pline Command	36
Exercise 8	39
Object Track (OTRACK)	40
Exercise 9	43
Polar Tracking (POLAR)	43
Exercise 10	46
Erase Command	47
Oops, Undo, and Redo commands	49
Redraw and Regen Commands	50
Exercise 11	51
Chapter Review	52
Chapter Review Answers	52

Chapter 3. How to Set Up Your Drawing — 53

What Are the Things to Think About	53
Step 1: Drawing Units	53
Step 2: Drawing Limits	55
Step 3: Layers	56
Changing an Object's Properties and Match Properties	65
Exercise 12	68
Workshop 1-A	69
Chapter Review	71
Chapter Review Answers	71

Chapter 4. A Few Good Construction Commands — 73

Introduction	73
Offset Command	74
Exercise 13	76
Fillet Command	77
Exercise 14	80
Chamfer Command	81
Exercise 15	83
Trim Command	84
Exercise 16	86
Extend Command	86
Exercise 17	89
Lengthen Command	89
Exercise 18	90
Workshop 2	91
Chapter Review	97
Chapter Review Answers	98

Chapter 5. Modifying Commands — 99

Introduction — 99
Selecting Objects — 100
Other Methods for Selecting Objects — 101
Move Command — 102
Exercise 19 — 103
Copy Command — 104
Exercise 20 — 105
Rotate Command — 106
Exercise 21 — 107
Scale Command — 108
Exercise 22 — 109
Array Command — 110
Exercise 23 — 112
Exercise 24 — 115
Mirror Command — 116
Exercise 25 — 117
Stretch Command — 117
Exercise 26 — 119
Break Command — 119
Exercise 27 — 120
Grips: Introduction — 121
Grips: The Five Commands — 122
Grips: Steps & Notes — 123
Grips and DYN — 123
Exercise 28 — 125
Chapter Review — 126
Chapter Review Answers — 127

Chapter 6. Dealing with Blocks — 129

What Are Blocks? — 129
Creating Blocks — 129
Workshop 3 — 131
Inserting Blocks — 133
Workshop 4 — 135
Exploding Blocks — 136
Sharing Data between AutoCAD Files Using Design Center (Blocks) — 137
Block Automatic Scaling — 140
Workshop 5 — 142
Tool Palettes: Introduction — 143
Creating Tool Palettes from Scratch — 145
Creating Tool Palettes Using Design Center — 147
Customizing Tool Palettes — 147
Workshop 6 — 150

Editing Blocks	151
Workshop 7	154
Chapter Review	155
Chapter Review Answers	155

Chapter 7. Hatching — 157

Hatching in AutoCAD	157
Bhatch Command: Selecting the Hatch Pattern	157
Bhatch Command: Selecting the Area to be Hatched	161
Bhatch Command: Preview the Hatch	162
Workshop 8	162
Bhatch Command: Options	164
Bhatch Command: Hatch Origin	166
Workshop 9	167
Bhatch Command: Advanced Features	169
Hatching Using Tool Palettes	171
Workshop 10	172
Gradient Command	172
How to Edit an Existing Hatch	176
Workshop 11	178
Chapter Review	179
Chapter Review Answers	179

Chapter 8. Text and Tables — 181

Introduction	181
Text Style	182
Workshop 12	185
DTEXT Command	186
MTEXT Command	187
Workshop 13	191
Editing Text	193
Editing Text Properties	193
Text and Grips	195
Spelling Check and Find and Replace	196
Workshop 14	197
Table Style	198
Workshop 15	202
Table Command	203
Workshop 16	205
Chapter Review	207
Chapter Review Answers	208

Chapter 9. Dimensioning your drawing — 209

Introduction	209
Dimension Types	210

Dimension Style: The First Step	213
Dimension Style: Lines Tab	214
Dimension Style: Symbols and Arrows Tab	217
Dimension Style: Text Tab	219
Dimension Style: Fit Tab	222
Dimension Style: Primary Units Tab	224
Dimension Style: Alternate Units Tab	226
Dimension Style: Tolerances Tab	227
Dimension Style: Creating a Child Style	229
Controlling Dimension Styles	230
Workshop 17	231
Dimensioning Commands: Linear	234
Dimensioning Commands: Aligned	235
Exercise 29	236
Dimensioning Commands: Arc Length	236
Dimensioning Commands: Ordinate	238
Dimensioning Commands: Radius	239
Dimensioning Commands: Jogged	240
Dimensioning Commands: Diameter	240
Dimensioning Commands: Angular	241
Exercise 30	242
Dimensioning Commands: Continue	243
Dimensioning Commands: Baseline	244
Exercise 31	244
Dimensioning Commands: Quick Dimension	245
Dimensioning Commands: Quick Leader	246
Dimensioning Commands: Center Mark	247
Dimension Blocks and Grips	248
Dimension Block Properties	249
Workshop 18	250
Notes	253
Chapter Review	254
Chapter Review Answers	254

Chapter 10. Plotting Your Drawing   255

Introduction	255
Model Space vs. Paper Space	255
Introduction to Layouts	256
Where Do I Find Layouts?	257
How to Create a New Layout	257
What is Page Setup Manager?	263
Workshop 19	266
Layouts and Viewports	267
Adding Viewports	268
Model Space and Paper Space Modes in Layouts	273

Modifying, Scaling, and Maximizing Viewports	274
Freezing Layers in Viewport	275
Workshop 20	276
Plot Style Tables: Introduction	279
Plot Style Tables: Color-dependent Plot Style Table	280
Plot Style Tables: Named Plot Style Table	284
Exercise 32	288
Plot Command	289
What Are DWF Files?	290
How Can We Produce Single-sheet and Multiple-sheet DWFs?	291
Workshop 21	296
Chapter Review	296
Chapter Review Answers	297
Appendix A. About the DVD	**299**
Appendix B. How to Create a Template File	**301**
Introduction	301
What Are the Elements to Include in a Template File?	301
How to Create a Template file	302
Appendix C. Inquiry Commands	**303**
Introduction	303
ID Command	303
DIST Command	304
AREA Command	304
LIST Command	305
Index	**307**

PREFACE

- AutoCAD is the de-facto drafting tool based on PCs since 1982. Millions of engineers, draftsmen, project managers, and engineering students use AutoCAD.
- This book will not teach the reader what engineering drafting is, and how to produce it. It is a prerequisite to know the science behind drafting.
- This book can be used in an ***instructor-led*** course, or as a ***teach-yourself*** guide:
  - As for the first option, the estimated time would be 3 days, 8 hours a day.
  - As for the second option, the reader can take it up at his/her convenience.
- There are 21 workshops, which will complete a full (small) project beginning with creating the project, up until the plotting. Solving all workshops will help to:
  - Simulate a real-life project from beginning to end, hence putting the reader in the practical mode.
  - Organize the information in a very logical order.
  - Learn the basic commands and functions in AutoCAD 2008.
- This book will cover the basic and intermediate levels of knowledge in AutoCAD 2008.

Chapter 1

INTRODUCTION TO AutoCAD 2008

In This Chapter

- What is AutoCAD?
- How to start AutoCAD
- Things you have to know about AutoCAD defaults
- Viewing commands
- Filing system in AutoCAD

WHAT IS AUTOCAD?

- AutoCAD is one of the first CAD (Computer Aided Design/Drafting) software applications.
- The first version of AutoCAD was released at the end of 1982, and it was designed to be used on PCs only.
- Since then, AutoCAD has enjoyed a wide user base all over the world.
- Users can draw both 2D drawings and 3D designs.
- There is another version of AutoCAD called AutoCAD LT, which is dedicated for 2D drafting only.
- In this book we will cover AutoCAD 2008.

HOW TO START AUTOCAD 2008

- There are two ways to start AutoCAD 2008:
  - While installing AutoCAD 2008, the installation program will create a shortcut on your desktop. If this is the case, simply double-click it.
  - From the Windows taskbar click **Start** / **All Programs** / **Autodesk** / **AutoCAD 2008** / **AutoCAD 2008**.

- AutoCAD will start with a new drawing file opened, just like the following:

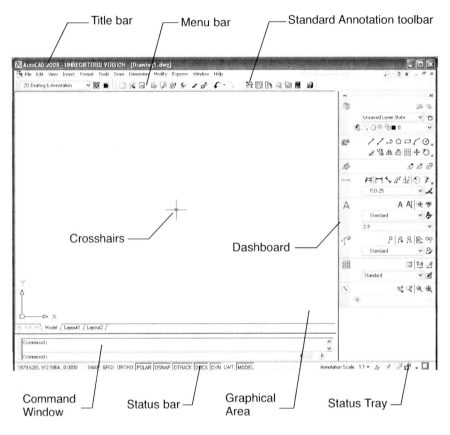

FIGURE 1-1

POINTS IN AUTOCAD

- Points are defined (and saved) in AutoCAD using the **Cartesian** coordinate system.
- The coordinates will look something like **3.25,5.45**, which is in the format of **X,Y**.
- The first and most traditional way of specifying points in AutoCAD is to type the coordinates as X,Y (we pronounce it as X comma Y). See the

below illustration:

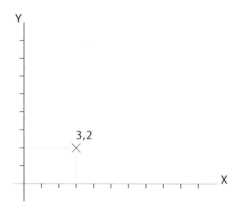

AUTOCAD DEFAULT SETTINGS

- Sign convention is: positive is up, and right.
- Angle convention is: positive is CounterClockWise (CCW) starting from the east (i.e., 0 angle). Check the following illustration:

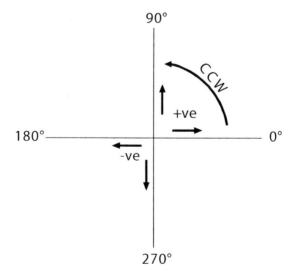

THINGS YOU HAVE TO KNOW ABOUT AUTOCAD

- A mouse is the primary input device:
  - Users can choose between a 2-button mouse and intelli-mouse (the one with a wheel).
  - The left button is always to "Select," "Pick," or "Click."
  - The right button means many things—see the next section.
- If you have an intelli-mouse (wheel mouse) you can use the wheel to:
  - Zoom in on your drawing by moving the wheel forward.
  - Zoom out of the drawing by moving the wheel backward.
  - Pan (i.e., moving through the drawing) by pressing the wheel and holding, then moving the mouse.
  - Zoom to the extents of your drawing by double-clicking the wheel.
- If you type an AutoCAD command or any input in the Command window you have to press the [Enter] key to execute it.
- [Enter] = [Spacebar] in AutoCAD.
- To repeat the last AutoCAD command, press [Enter] or [Spacebar].
- To cancel any AutoCAD command, press [Esc].

HOW TO ISSUE COMMANDS IN AUTOCAD

- There are several ways to issue commands in AutoCAD:
  - Type the command in the Command window (this is the oldest method, only old veterans of AutoCAD use this).
  - Use pull-down menus.
  - Use toolbars.
  - Use Dashboard (this is the new method of issuing commands for 2D, in AutoCAD 2007 it was only for 3D commands).

DYN AND TYPING AUTOCAD COMMANDS

- If the DYN button on the status bar is on, then anything you type in the Command window will appear on the screen beside the AutoCAD cursor.
- For example, if you type the word LINE, here is how it will look on the screen:

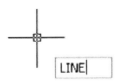

- When you press [Enter], the following will appear:

- You can now type the X coordinate, and then press the [Tab] key, and then input the Y coordinate.
- We will cover **DYN** in the coming chapter.

DRAWING LIMITS

- AutoCAD offers the user an infinite drawing sheet on all sides.
- When you open AutoCAD, your viewpoint will be at 0,0,1.
- You are looking at the XY plane, using a camera's lens; hence you will see part of your infinite drawing sheet. This part is called the **limits**. See the illustration below:

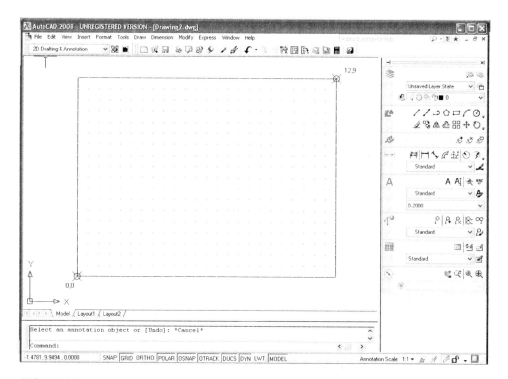

FIGURE 1-2

- In the above example, you will see that the limits of the drawing are from 0,0 (lower left corner) to 12,9 (upper right corner), which means this is the working area you should place your work in.
- We will learn how to change the limits in the coming chapters.

AUTOCAD UNITS AND AUTOCAD SPACES

- One of the vague facts about AutoCAD is that it doesn't deal with a certain length unit while drafting. So let's clarify the following points:
  - AutoCAD deals with AutoCAD units.
  - AutoCAD units can be anything you want. They can be meters, centimeters, millimeters, inches, or feet.
  - All of these assumptions are right as long as you remember your assumption later on, and you are consistent in both X and Y.
- Also, there are two spaces in AutoCAD, **Model Space** and **Paper Space (Layouts)**, if you are looking at the lower left corner of the graphical screen you will see the following:

[Model / Layout1 / Layout2]

- In Model Space you will create, modify, and annotate.
- Once you start thinking about making a hardcopy of your drawing file, go to the Paper Space (Layouts) so you can prepare your page setup.
- This is the moment you need to remember, *"What is my unit assumption?"* so you can scale your drawing to the proper scale.
- We will discuss printing later.

RIGHT-CLICKING

- Starting with AutoCAD 2000, right-click definitions changed dramatically.
- In each place of the screen there is a reason for right-clicking. We will discuss a few of these:
  - If you right-click a toolbar, a shortcut menu will appear showing all the toolbars available in AutoCAD so you can show/hide any toolbar.

INTRODUCTION TO AutoCAD 2008   7

- If you right-click the Command window, a shortcut menu will appear to help you repeat commands you already used, such as:

- Other right-clicking definitions will come during the discussion of commands in AutoCAD 2008.

VIEWING COMMANDS

- We discussed previously in this chapter the benefits of the new intelli-mouse wheel for zooming in, zooming out, and panning.
- If you don't have an intelli-mouse, you still can zoom in, zoom out, and pan using the zooming and panning commands.
- From menus you can use **View/Zoom**, and **View/Pan**, there are lots of options to choose from.
- From the **Standard** toolbar, you will find the following buttons:
  - **Pan Realtime** (just like pressing and holding the wheel).

  - **Zoom Realtime**, click the left button of the mouse, and hold. If you move forward, you are zooming in, if you move backward you are zooming out.
  - **Zoom Window**, to specify a rectangle. By specifying two opposite corners, whatever inside the rectangle will look larger.
  - **Zoom Previous**, to restore the previous view, up to the last ten views.

- From the **Zoom** toolbar, use the following zooms:
  - **Zoom Dynamic**, to utilize this command, you better do Zoom Window first. You will see the whole drawing and your current place (it will be shown as a green dotted line). Go to the new location and press [Enter].
  - **Zoom Scale**, AutoCAD will ask you to input a scale factor, so type in a number. If it is less than 1, you will see the drawing smaller. If the scale factor is greater than 1, you will see the drawing larger. If you put the letter x after the number (e.g., 2x) the scale will be relative to the current view.
  - **Zoom Center**, to specify a new center point for the zooming, along with a new height.
  - **Zoom Object**, to zoom to certain selected objects. AutoCAD will ask you to select objects. The selected objects will fill the screen.
  - **Zoom In**, is not really a zoom option, but a programmed option. It equals **Zoom/Scale**, with a scale factor of 2x.
  - **Zoom Out**, just like Zoom In, but with a zoom factor of 0.5x.

- **Zoom All**, to zoom to all objects in the drawing or limits, whichever is greater.

- **Zoom Extents**, to zoom to all objects.

CREATE A NEW FILE

- To create a new file based on a premade template, do one of the following:
  - From the **Standard Annotation** toolbar, click **QNew** button.
  - From menus select **File/New**.
  - Press **Ctrl+N**.
  - Type the **new** command in the Command window.
- The following dialog box will appear:

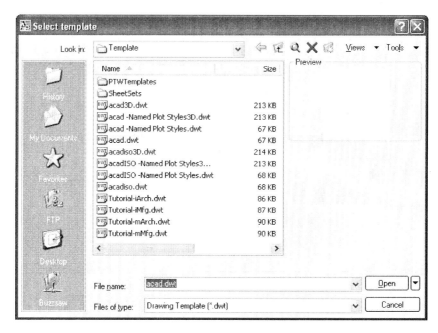

FIGURE 1-3

- This dialog box will allow you to select the desired template.
- AutoCAD template files have the extension *.dwt*.
- AutoCAD 2008 has lots of premade templates that you can use, or you can create your own templates.

- For now we will use *acad.dwt* to help us with some of our exercises and workshops.

- After you select the desired template file, click the **Open** button, a new file will open.

OPEN AN EXISTING FILE (SINGLE FILE)

- To open an existing file for further editing, do one of the following:
  - From the **Standard Annotation** toolbar, click the **Open** button.
  - From menus select **File/Open**.
  - Press **Ctrl+O**.
  - Type the **open** command in the Command window.
- The following dialog box will appear:

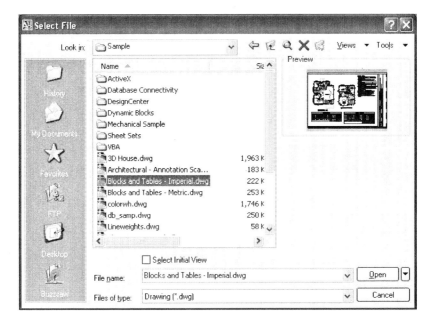

FIGURE 1-4

- Specify the hard drive and the folder your file resides in.
- AutoCAD drawing files have the extension *.dwg*.
- Click on the desired file. You will see a preview to the right of the dialog box, so you can make sure you are opening the right file.

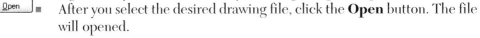
- After you select the desired drawing file, click the **Open** button. The file will opened.

OPEN AN EXISTING FILE (MULTIPLE FILES)

- You can open as many files as you wish in AutoCAD. Your limit is defined by your machines capabilities.
- Use the same command for opening a file we used on the previous page. This time, instead of selecting a single file, click the first file, then using the [Ctrl] key, select more files, just like the following:

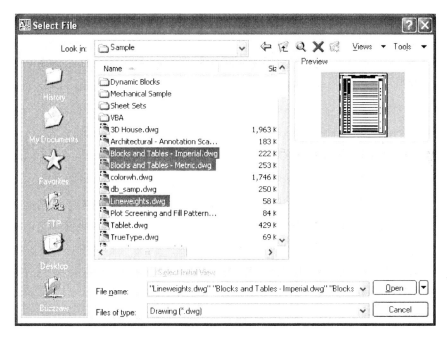

FIGURE 1-5

- Once you are done, click the **Open** button. The files will open one by one.
- You will see only the last file. To view the rest of the files; from menus select **Window**, you will see the following:

- As you can see, only one file has a (✓) to the left of its name. If you want to view another file, simply click the name and it will appear on the screen.
- To tile the files vertically, from menus select **Window/Tile Vertically**. This is what you will get:

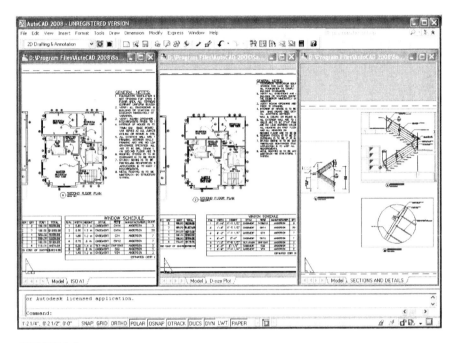

FIGURE 1-6

- To tile the files horizontally, from menus select **Window/Tile Horizontally**. This is what you will get:
- To switch between files, use either **Ctrl+F6** or **Ctrl+Tab**.
- To close all files using a single command, from menus select **Window/Close All**.
- To close one file at a time, use one of the following:
  - From menus select **File/Close**.
  - From menus select **Window/Close**.
- You can open the last used file by clicking the **File** menu.
  - Above the **Exit** command you will see a list of files.
  - Select one of these and it will open.

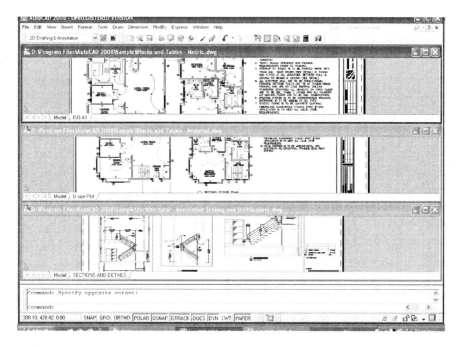

FIGURE 1-7

SAVE AND SAVE AS FILES

- After you are done with your design you can save your work by using one of the following methods:
  - From the **Standard Annotation** toolbar, click the **Save** button.
  - From menus select **File/Save**.
  - Press **Ctrl+S**.
  - Type the **qsave** command in the Command window.
- If this is the *first* time to save your work, the following dialog box will appear:
- Type in the desired filename and click the **Save** button.
- If you saved your file before, this command will save it under the same name.
- To save your file under another name, do one of the following:
  - From menus select **File/Save As**.
  - Press **Ctrl+Shift+S**.
  - Type **save as**, or **save** in the Command window.
- The above dialog box will appear, so you can type the new filename and click the **Save** button.

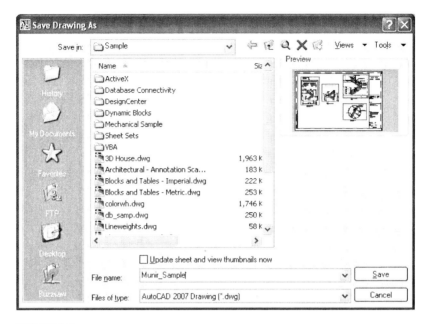

FIGURE 1-8

EXITING AUTOCAD

- To exit AutoCAD, use one of the following methods:
    - From menus select **File/Exit**.
    - Press **Ctrl+Q**.
    - Type **quit** or **exit** in the Command window.

- Whether you are closing a file(s) or exiting AutoCAD, AutoCAD will perform a check on the current file(s). If any change took place from the time of opening until the time of closing it will show the following message:

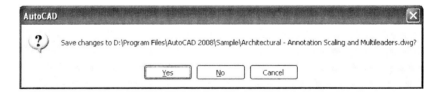

- If you want to save the changes, click **Yes**.
- If you want to discard these changes, click **No**.
- If you want to cancel the whole command (closing or exiting), click **Cancel**.

NOTES:

INTRODUCING AUTOCAD 2008

 Exercise 1

1. Start AutoCAD 2008
2. Select **File/Open**
3. Go to the **Samples** folder under the AutoCAD 2008 folder and open 3 files of your choice.
4. Using the Window menu, switch between files, then tile them vertically. Using Ctrl+F6 (or Tab), switch between tiled files.
5. Using one of the files opened, and using the wheel on your mouse or any of the zoom commands you know, zoom in on the drawing, zoom out (using different methods), and pan.
6. Using the **Window/ Close All** menu, close all files without saving.

CHAPTER REVIEW

1. You can close all opened files using one command:
    a. True, using File/Close All
    b. True, using File/Close
    c. True, using Window/Close All
    d. False
2. CAD means _____
3. AutoCAD has two spaces; Model Space and Paper Space:
    a. You draw on Model Space and print from Paper Space.
    b. You draw in Paper Space, and print from Model Space.
    c. There is only one space in AutoCAD.
    d. Model Space is only for 3D design.
4. Positive angles are starting from the NORTH:
    a. True
    b. False
5. AutoCAD is one of few pieces of software that will allow the user to:
    a. Connect through the Internet
    b. Type commands using the keyboard
    c. Accept Cartesian coordinates
    d. Positive angles are CCW
6. You can use Ctrl + _____, or Ctrl + _____ to switch between opened files.

CHAPTER REVIEW ANSWERS

1. c
2. Computer Aided Design/Drafting
3. a
4. b
5. b
6. F6, Tab

Chapter 2

DRAFTING USING AUTOCAD 2008

In This Chapter

- Line command and two precision methods
- Arc and Circle commands
- Object Snap (OSNAP)
- Pline command
- Object Tracking (OTRACK)
- Polar Tracking (POLAR)
- Erase and basic selecting methods

INTRODUCTION

- There are two important things in drafting:
  - Precision
  - Speed
- You always want to finish your drawing as fast as possible, yet you don't want to undermine your drawing's precision.
- We always put precision before speed, as it is easier to teach you how to speed up your creation process if you are slow. It is nearly impossible to teach somebody the good techniques of accuracy.
- In this chapter we will tackle many commands, but the most important issue here is drafting with precision.

LINE COMMAND

- To draw segments of straight lines.
- There are lots of methods to help us draw precise shapes using the Line command, which will be discussed later. As for now we will use

the only method we know—typing the coordinates in the Command window.
- To issue the Line command do one of the following:
  - From the dashboard and using the **2D Draw** panel, click the **Line** button.
  - From menus, select **Draw/Line**.
  - Type **line** (or **l**) at the Command window.
- The following prompts will appear:

  ```
  Specify first point: (type in the coordinates of
  the first point)
  Specify next point or [Undo]: (type in the coor-
  dinates of the second point)
  Specify next point or [Undo]: (type in the coor-
  dinates of the third point)
  Specify next point or [Close/Undo]: (type in the
  coordinates of the fourth point)
  ```

- At any moment you can use the **Undo** option to undo the last specified point, hence the last specified segment.
- After you draw two segments the **Close** option will be available to connect the last point to the first point and end the command.
- Other ways of ending the command are to press [Enter] or the spacebar.
- Also, if you press [Esc], Line command would end.
- If you are in Line command and you right-click, you will get the following shortcut menu, which is identical to the command prompts.

DYNAMIC DRAFTING

- We saw in Lesson 1 the **DYN** button in the status bar and how it shows anything you type on your keyboard, starting from issuing the command, along with the entry of the coordinates.

- **DYN** has another very important feature with Line command, which is showing the length and the angle of the line to be drawn (the angle is measured from the east and incremented by 1 degree).
- Take a look at the following example:

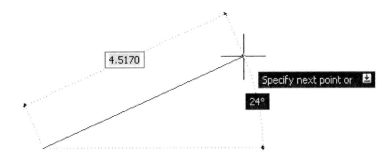

- As you can see, once you start Line command and you specify the first point, move your mouse to the right and you will see the length of your proposed line and its angle measured from the *east*.
- Specify the length, press the [Tab] key, and then type the angle.

DRAWING LINES – FIRST METHOD

 Exercise 2

1. Start AutoCAD 2008.
2. Open the file **Exercise_02.dwg**.
3. Draw the following lines using the **LINE** command and **DYN**:

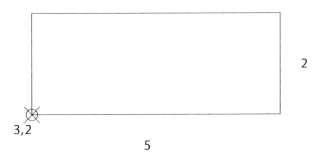

 Start LINE command, then type the coordinates of the first point. While DYN is on, specify the length, press [Tab], then specify the angle, do that for the other line segments.

4. Save the file and close it.

PRECISION METHOD 1: SNAP AND GRID

- As you can see, the only method we used to specify points in the XY plane was to type the coordinates using the keyboard using DYN.
- We are doing this because we can't depend on the mouse to specify precise points because the mouse is free to move anywhere in the XY plane.
- In order to depend on the mouse to work precisely, we need to control its movement.
- **Snap** is the only tool in AutoCAD that can help us control the movement of the mouse.

- Using the status bar, click the **SNAP** button on.
- Now move on to the Graphical area and watch the mouse jump to exact points.
- **Grid** would show a grid of points on the screen to simulate a paper with a grid of lines, which is used in drawing diagrams (these points are not real points).
- Grid by itself is not accurate, but is considered a helping tool when used with Snap.

- Using the status bar, click the **GRID** button on.
- You can now see the points displayed on the screen.
- If you think the default values for either Snap or Grid do not satisfy your need, simply right-click one of the two buttons and the following shortcut menu will appear:

- Select **Settings** and the following dialog box will appear:
- By default, **Snap X spacing** and **Snap Y spacing** are equal. Also, **Grid X spacing** and **Grid Y spacing** are equal. If you want this to continue make sure that the checkbox **Equal X and Y spacing** is always on.
- By default, if you are working with 2D, you will only see Grid dots. But if you work with 3D you will see Grid lines, hence set the **Major line spacing**.
- Also, all the settings of **Grid behavior** are meant for 3D drawing.
- Make sure that **Snap type** is **Grid snap** (we will discuss **PolarSnap** in the coming pages). If you are creating a 2D drawing, then select the

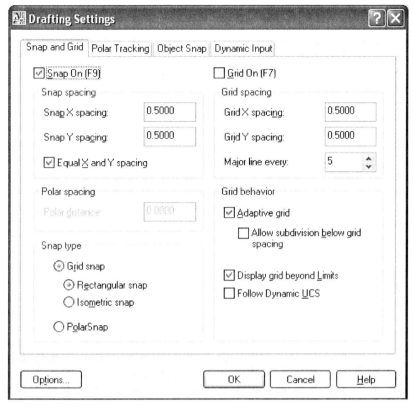

FIGURE 2-1

Rectangular snap. If you are creating a 3D drawing select **Isometric snap**.
- If you want Grid to follow Snap, set the two values of Grid to zero.

- You can use function keys to turn on both Snap and Grid:
  - F9 = Snap on/off
  - F7 = Grid on/off

SNAP AND GRID

Exercise 3

1. Start AutoCAD 2008.
2. Open the file **Exercise_03.dwg**.

3. Using **DYN**, **SNAP**, and **GRID**, draw the following lines without typing any coordinates on the keyboard (don't draw the dimensions):

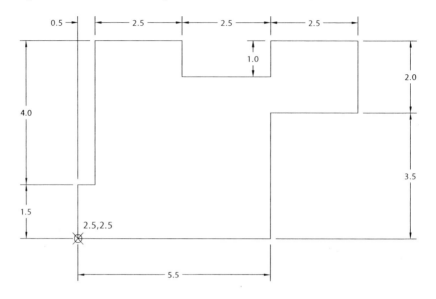

4. Save the file and close it.

Change the Snap X spacing to 0.5 first and set the Grid values to 0. Switch both Snap and Grid on and draw the lines as required.

PRECISION METHOD 2: DIRECT DISTANCE ENTRY AND ORTHO

- If we know that lines in AutoCAD are ***vectors***, which means we need to specify a length and an angle to successfully draw them, we will appreciate this method.
- **Ortho** is a tool that will force the cursor to always give us orthogonal angles (i.e., 0, 90, 180, and 270).
- Direct Distance Entry is a very handy tool in drafting; if the mouse is already directed toward an angle just type in the distance and press [Enter].
- Combining the two tools will allow us to draw lines with precise lengths and angles.

- Do the following steps:

  - From the status bar, click the **ORTHO** button on.
  - Start Line command.
  - Specify the first point (either by typing or using Snap).
  - Move the mouse to the right, up, left, and down, and see how it gives us only orthogonal angles.
  - Use the desired angle, type in the distance, and press [Enter].
  - Continue with other segments using the same method.
- You can also use Direct Distance Entry with **DYN** using the same method mentioned above.

DIRECT DISTANCE ENTRY AND ORTHO

 Exercise 4

1. Start AutoCAD 2008.
2. Open file **Exercise_04.dwg**.
3. Using **ORTHO** and Direct Distance Entry draw the shape below (without dimensions):

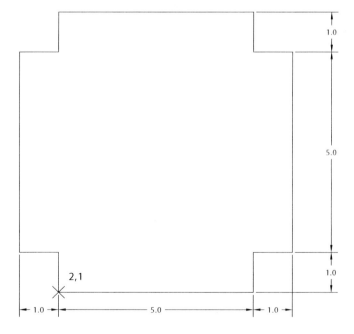

4. Save the file and close it.

ARC COMMAND

- The Arc command is used to draw a circular arc (arc part of a circle).
- Check the following illustration:

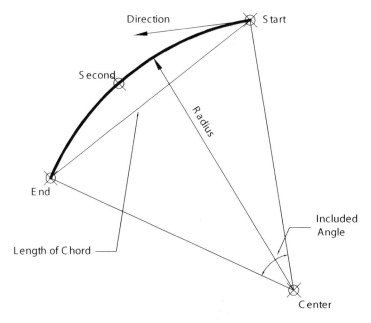

- The details that AutoCAD recognizes about an arc are:
  - Start point
  - Second point (not necessarily the mid point)
  - Endpoint
  - Center point
  - Radius
  - Length of Chord
  - Included Angle (angle between Start-Center-End)
  - Direction (the tangent passes through the Start point)
- AutoCAD needs only three pieces of information to draw an arc, but not any three.
- AutoCAD will start asking you to make your first input choosing between Start point or Center point, and based on the choice it will ask you to specify the second piece of information, and so on.
- To make it easier for users, AutoCAD designers programmed the ten possible methods to draw an arc in the menus.

DRAFTING USING AutoCAD 2008   27

- From menus, select **Draw/Arc**, the following menu will appear:

- Before you start, specify the method desired, and then select it from the menu. AutoCAD will take it from there.
- Always think counterclockwise when specifying points.

- From the dashboard and using **2D Draw** panel, click the **Arc** button— or you can type **arc** or **a** at the command window, but this method can be a little bit tough on novice users.
- The following prompt will appear:

```
Specify start point of arc or [Center]: (type in
the coordinates of the start point, or type the
letter c to specify the center point)
Specify second point of arc or [Center/End]: (if
you specified the first point, you are allowed to
input either second point, center point, or radius)
Specify end point of arc: (if you specified the
```

second point, you will be asked to specify the end point)

- Also, while you are using this method, you can use the right-click.
- After you specify the Start point, specify the second point or right-click, the following shortcut menu will appear:

- Select either Center or End. If you select Center, specify the center point. Now, either you will specify the Endpoint or right-click, the following shortcut menu will appear:

DRAWING ARCS

Exercise 5

1. Start AutoCAD 2008.
2. Open the file **Exericise_05.dwg**.
3. Turn on **SNAP** and **GRID**.

4. Draw the four arcs as illustrated, using any of the methods you learned:

5. Save the file and close it.

CIRCLE COMMAND

- There are five different methods to draw a circle in AutoCAD (the sixth method is not really genuine but rather programmed).
- You have to know the Center of the circle to use the first two methods. They are:

Center/Radius   Center/Diameter

- To use the third method you have to know any three points on the parameter of the circle:

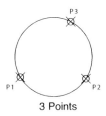

3 Points

- To use the fourth method, specify two points on the parameter of the circle with the distance between them equal to the diameter:

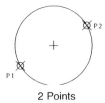

2 Points

- To use the fifth method you should have two objects already drawn, so we can consider them as tangents, then specify a radius:

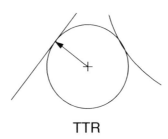

TTR

- The sixth method is not part of AutoCAD options if you used the Command window, but instead it is programmed and included in the menus only.
- To use it, specify three objects as tangents (the original method used here is the 3-points method).
- Just like the Arc command, AutoCAD designers programmed the 6 methods in the menus.
- From menus, select **Draw/Circle** and the following will appear:

- Before you start, specify the method desired and then select it from the menu. AutoCAD will take it from there.
- As in the **Arc** command there are other ways to reach the **Circle** command, which is more difficult for novice users.
- From the dashboard and using **2D Draw** panel, click the **Circle** button—or you can type **circle** or **c** at the Command window.
- The following prompt will appear:

```
Specify center point for circle or [3P/2P/Ttr (tan
tan radius)]: (Type in the circle center, or type
one of the capital letters to start the other
methods)
Specify radius of circle or [Diameter]: (If you
specified the center, AutoCAD will ask you to
specify either radius, or type d, to specify the
diameter)
```

- Also, while you are using this method you can use the right-click.
- Once you issue the command, either you will specify the center point or right-click, the following shortcut menu will appear:

- Select the desired method.

DRAWING A CIRCLE

Exercise 6

1. Start AutoCAD 2008.
2. Open the file **Exericise_06.dwg**.
3. Make sure that **SNAP** and **GRID** are on.
4. Draw the four circles as illustrated (Radius = 0.5).

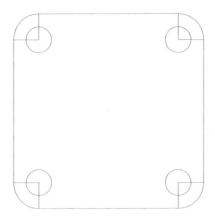

5. Save the file and close it.

PRECISION METHOD 3: OBJECT SNAP (OSNAP)

- AutoCAD keeps a full record of each object in each drawing.
- Object Snap is a tool that helps you utilize this record when you need to specify points on objects already drawn.
- Example:
  - Assume we have the following two lines:

  - We have no information about these two lines.
  - We are asked to draw a precise line from the mid of the upper line to the end of the lower line.
  - The command to draw is **line**. AutoCAD asks us to specify the first point. We type **mid** and press [Enter] or spacebar, then go directly to the upper line where a yellow triangle appears. We click on it.

- AutoCAD asks us to specify the next point. We type **end** and press [Enter] or spacebar, then go directly to the lower line where a yellow rectangle appears. We click on it and then press [Enter] to end the command.

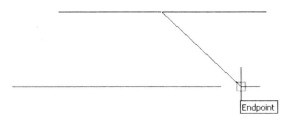

- Mission accomplished.
- Some Object Snaps buttons are:

- To catch the Endpoint of an object.

- To catch the Midpoint of an object.

- To catch the Intersection of two objects.

- To catch the Center of an arc, or a circle.

- To catch the Quadrant of an arc, or a circle.

- To catch the Tangent of an arc, or a circle.

- To catch the Perpendicular point on an object.

- To catch a point on an object Nearest to your click point.

- We will discuss more and more about other Object Snaps when we learn more commands.

- There are four ways to use the Object Snaps whenever you are asked to specify a point. These are:

Typing
- Type the first three letters of the desired OSNAP such as: end, mid, cen, qua, int, per, tan, or nea.

Shit+Right-click
- Hold the [Shift] key, and right-click. The following shortcut menu will appear. Select the desired OSNAP.

Toolbar
- Right-click on any toolbar and select the **Object Snap** toolbar. The following toolbar will appear. Click the desired OSNAP whenever it's needed.

Running OSNAP
- This method is the most practical method of all of the above.

- From the status bar, click the **OSNAP** button on (you can click F3).
- To set the desired OSNAP that we want to be running with us, right-click on the **OSNAP** button. The following shortcut menu will appear:

- Select **Settings** and the following dialog box will appear:

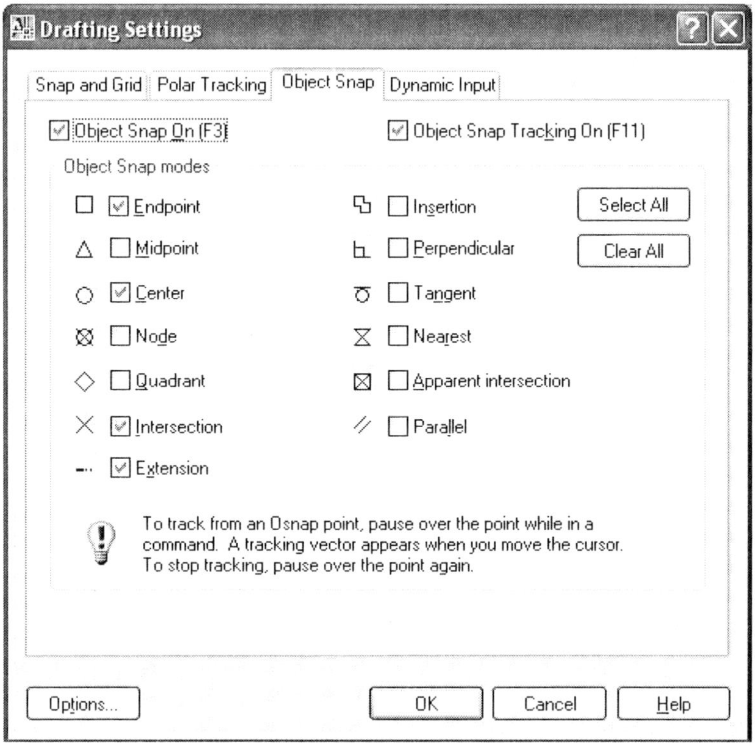

FIGURE 2-2

- Switch on the desired OSNAP and click OK.
- Now each and everytime you are asked for a point, all you have to do is to go to the place you want and once the OSNAP shape appears, AutoCAD captures the desired point.

DRAWING USING OSNAP

Exercise 7

1. Start AutoCAD 2008.
2. Open the file **Exercise_07.dwg**.

3. Draw the following shape (without dimensions), using the following steps:

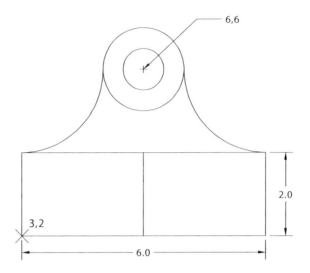

a. Starting from point 3,2 and using Direct Distance Entry and ORTHO, draw the four lines, which represent the large rectangle (6 x 2).
b. Using the Circle command method Center/Radius, draw the large circle using center = 6,6 and radius = 1.
c. Set OSNAP to give you Endpoint, Midpoint, Center, and Quadrant.
d. Draw the small circle using the Center of the large circle and radius = 0.5.
e. Draw a vertical line from the midpoint of the upper horizontal line to the midpoint of the lower horizontal line.
f. To draw the left arc, use the method Start/End/Angle. Start from the upper left corner of the rectangle using the endpoint, then click the left quadrant of the large circle, then type 90, and press [Enter].
g. To draw the right arc, use the method Start/End/Angle. Start from the upper right corner of the rectangle using the endpoint, then click the right quadrant of the large circle, then type -90, and press [Enter]
4. Save the file and close it.

PLINE COMMAND

- Pline means Polyline. Poly means many, so if you exchange 'many' with 'poly,' the new name would be many lines.

- At the beginning we will make a small comparison study between Line command and Polyline command.

Line	Polyline
Each segment is an object	One object with vertices
Lines only	Lines and Arcs
No width	Variable Width for start and finish

- As you can see from the comparison there are mainly three differences between the two commands.
- To issue a Pline command do one of the following:
    - From the dashboard and using **2D Draw** panel, click the **Polyline** button.
    - From menus, select **Draw/Polyline**.
    - Type **pline** (or **pl**) in the Command window.

- The following prompt will appear:

```
Specify start point: (just like you did with line
command)
Current line-width is 0.9000
Specify next point or [Arc/Halfwidth/Length/Undo/
Width]: (specify the next point or choose one of
the options available)
```

- After you specify the first point, Polyline will report to you the current Polyline width (in our example it is 0.90), then it will ask you to specify the next point. You can use all the methods we learned in Line command.

- If you don't want to specify the second point, you can choose from the following options:

Arc
- By default Polyline command will draw lines.
- You can change the mode to draw arcs by selecting this option. The following prompt will appear:

  ```
  Specify endpoint of arc or
  [Angle/CEnter/CLose/Direction/Halfwidth/Line/Radi
  us/Second pt/Undo/Width]:
  ```

- We learned in Arc command that AutoCAD needs three pieces of information to draw an arc.
- AutoCAD already knows the start point of the arc, which is the start point of the polyline, or the endpoint of the last line segment.
- AutoCAD will make a certain assumption that the user has the right to accept or reject. This assumption is that the Direction of the arc will be the same angle as the last line segment.
- If you accept this assumption, then AutoCAD will ask you to specify the endpoint of the arc.
- If you reject this assumption, then specify the second piece of information from the following:
  - Angle of Arc, then Center or Radius.
  - Center, then Angle or Length.
  - Direction, then End.
  - Radius, then End or Angle.
  - Second, then End.

Half-width
- The first method to specify the width of Polyline.
- Specify the half-width of the polyline from Center to one of its edges, something like the following:

- When you select this option, AutoCAD will show the following prompt:

  ```
  Specify starting half-width <1.0000>:
  Specify ending half-width <1.0000>:
  ```

- In our above example, the half-width is 1.0 for both the start and end.

Length
- In polyline command, if you draw an arc, then switch to Line to draw a line segment, and you want the line to be tangent to the arc, then select this option.

| | This option will assume the angle to be the same as the last segment, hence will ask you only for the length. The following prompt will appear:

```
Specify length of line:
```

Width | Same as half-width, but instead you have to input the full width. See the illustration below:

 | The **Undo** and **Close** options are the same options in Line command.
| If you choose to close in the Arc option it will close the shape by an arc.

DRAWING POLYLINES

 Exercise 8

1. Start AutoCAD 2008.
2. Open the file **Exercise_08.dwg**.
3. Using **ORTHO** and Direct Distance Entry draw the following shape (without dimensions) using the **Pline** command with Width = 0.1.

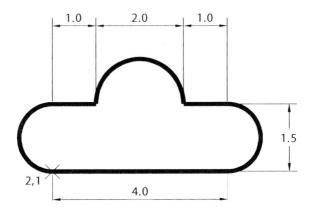

4. Save the file and close it.

 | Take care of the following tips:
- In order to draw the large arc, use Angle = 180.
- Before you draw the last arc, change the mode to **arc** and select **CLose**.

OBJECT TRACK (OTRACK)

- If you have a rectangle, and you want to draw a circle whose center will coincide with the center of the rectangle exactly. OTRACK will help you do this without drawing any new objects to facilitate specifying the exact points.
- OTRACK uses OSNAPs of existing objects to *steal* the coordinates of the new point.
- From the Status bar, click the **OTRACK** button on (you can press F11).
- Make sure that OSNAP is also on, as OTRACK alone won't do any thing.
- Let's take an example where we will use two points to specify one point:
  - Assume we have the following rectangle.

Example of two-point OTRACK

- Make sure that OSNAP and OTRACK are both turned on. Make sure that Midpoint in OSNAP is turned on.
- Start the Circle command, which will ask you to specify the center point.
- Go to the upper (or lower) horizontal line and move to the midpoint and stay there for a couple of seconds, then move up or down. You will see an infinite line extending in both directions, just like the following:

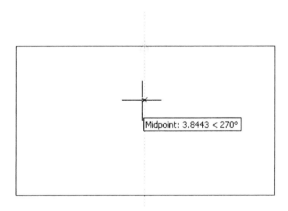

- Go to the right (or left) vertical line and move to the midpoint and stay there for a couple of seconds, then move right or left. You will see an infinite line extending in both directions, just like the following:

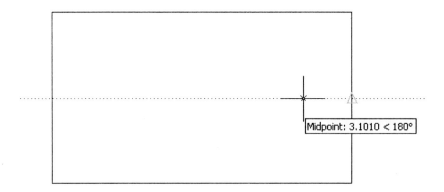

- Now go to where you think the two infinite lines should intersect.

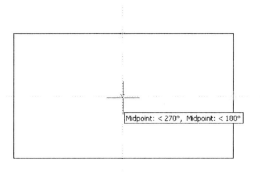

- Once you see the two infinite lines, click. You have now specified the center point of the circle. Next type in the radius of the circle.

Example of One-point OTRACK

- Let's take another example, but this time we will use one point to specify one point:
- Continue with the same shape we did in the last example:

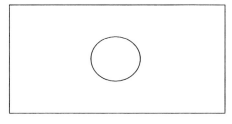

- Start the Circle command, which will ask you to specify the center point.
- Make sure both OSNAP and OTRACK are turned on, and also turn on Center in OSNAP.
- Go to the center point of the existing circle and stay there for a couple of seconds, then move to the right. An infinite line will appear:

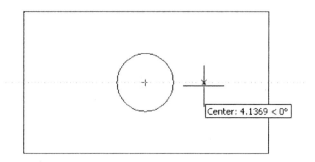

- Type 5 (or any distance) and press [Enter]. The center of the new circle will be specified, then type in the radius.
- This is what you will get:

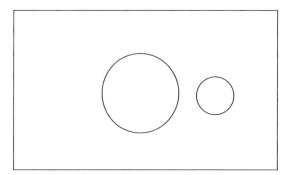

NOTE ■ If you stayed over a point for a couple of seconds to produce an infinite line, then you discovered that this isn't the desired point. Simply go to the same point again and stay over it again for a couple of seconds, it will be disabled.

DRAWING USING OSNAP AND OTRACK

 Exercise 9

1. Start AutoCAD 2008.
2. Open the file **Exercise_09.dwg**.
3. Turn **OSNAP** on, and set the following OSNAPs on: Endpoint, Midpoint, and Center.
4. Turn on **OTRACK**.
5. Draw the four circles while specifying the center using OSNAP and OTRACK.

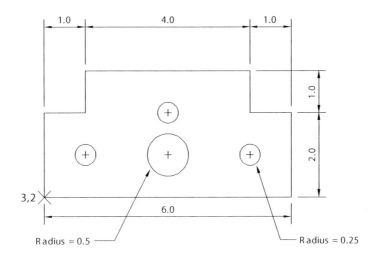

6. Save the file and close it.

POLAR TRACKING (POLAR)

- We learned that we can force the cursor to give us four orthogonal angles (0, 90, 180, 270) using ORTHO.
- If we want other angles such as 30, and its multiples, or 60 and its multiples, ORTHO wouldn't help us.
- For this purpose AutoCAD designers invented another powerful tool called Polar Tracking.
- Polar Tracking allows you to have rays starting from your current point pointing toward angles such as 30, 60, 90, 120, and so on. You can use Direct Distance Entry, just like we did with ORTHO.

POLAR
Increment angle

- From the Status bar, click the **POLAR** button on (you can press F10).
- To set the angles to be used, from the Status bar right-click on the **POLAR** button and select **Settings**. The following dialog box will appear:

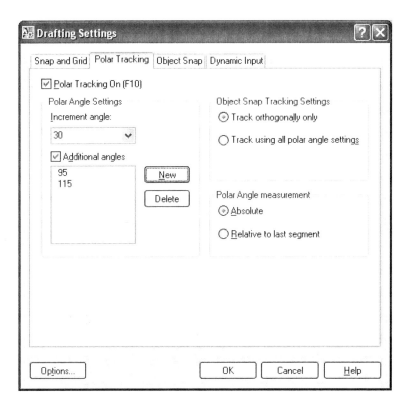

FIGURE 2-3

- Under **Polar Angle Settings**, select the **Increment angle** pop-uplist. You will find predefined angles, select the desired angle.
- If the desired angle is not among the list, simply type your own angle.
- Based on the above example, the user will have rays in angles 0, 30, 60, 90, 120, . . . , etc.

Additional Angles

- Sometimes, in the design process, you will need some odd angles that the Increment angle can't give you, such as 95 or 115.
- The option **Additional angles** will help you set these odd angles.
- Using the same dialog box, click the **Additional angles** checkbox on.
- Click the **New** button, and type in the angle.

- To delete an existing additional angle, select it, and click the **Delete** button.
- You will have something similar to the following:

Polar Snap
- AutoCAD will not give the multiples of the additional angles.
- We saw previously the command SNAP, which helped us to specify exact points on the XY plane using the mouse.
- The SNAP command can help us only along the X-axis (+ve and –ve), and along the Y-axis (+ve and –ve).
- If you want to snap to point along the ray produced by POLAR, you have to change the type of the SNAP from **Grid Snap** to **Polar Snap**.
- From the Status bar, switch on the **SNAP**. Right-click on the **SNAP** button and select **Settings**. Under **Snap type**, select **Polar Snap** instead of **Grid snap**, just like the following:

- Now set the **Polar spacing** value, just like the following:

Example
- We want to draw the following shape:

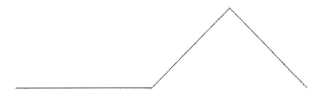

- Let's assume we set the Increment angle to 45, and we changed the type of SNAP to Polar Snap with Snap distance = 0.5. To draft using Polar Tracking follow the following steps:
  - Start the Line command, then specify a starting point.
  - Move to the right until you see a ray coming out, read the distance and the angle. When you reach your distance click to specify a point, just like the following:

  - Move the cursor toward the angle 45 until you see the ray. Now move the mouse to the desired distance and click:

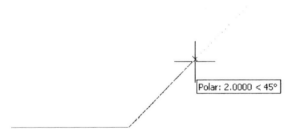

  - Move the cursor toward the angle 315 until you see the ray. Now move the mouse to the desired distance and click:

DRAWING USING POLAR TRACKING

 Exercise 10

1. Start AutoCAD 2008.
2. Open the file **Exercise_10.dwg**.

3. Switch both POLAR and SNAP on, and make the following settings:
   a. Increment angle = 30
   b. Additional angles = 135
   c. Polar distance = 0.5
4. Draw the following shape (without dimensions) starting from 3,2:

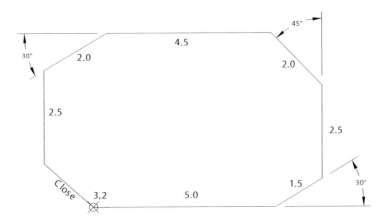

5. Save the file and close it.

ERASE COMMAND

- **Erase** will be the first modifying command to be discussed in this book.
- The only purpose of this command is to delete any object you select.
- You can reach this command using one of the following methods:
  - From the **Modify** toolbar, click the **Erase** button.
  - From menus select **Modify/Erase**.
  - Type **erase** (or **e**) in the Command window.
- Using any of the above methods, AutoCAD will prompt:

  ```
  Select objects:
  ```

- Once this prompt appears the cursor will change to a pick box:

- Basically you can do three things with the pick box:
  - Touch an object and click to select it.
  - Go to an empty place, click and go to the right, this will open a **Window**.
  - Go to an empty place, click and go to the left, this will open a **Crossing**.

Window
- A window is a rectangle specified by two opposite corners. The first corner is where you click on the empty place. Release you hand, go to a suitable second place, and click for the second corner.
- Whatever is inside the rectangle FULLY will be selected. If any part (even a tiny part) is outside the rectangle it will not be selected.

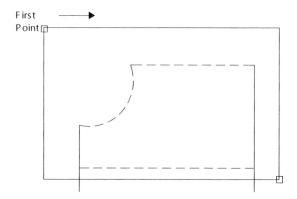

Crossing
- A crossing is just like a window, except that whatever is inside it will be selected, AND whatever it touches as well.

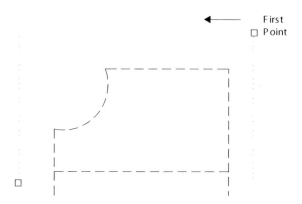

- These three methods can be used with almost all the modifying commands and not only the **Erase** command.

- The Select Objects prompt is repetitive. You always need to finish by pressing [Enter] or right-click.
- Other ways to erase objects are:
  - Without issuing any command, click on the object(s) desired, then press the [Del] key.
  - Without issuing any command, click on the object(s) desired, then right-click. The following shortcut menu will appear, select **Erase**.

OOPS, UNDO, AND REDO COMMANDS

- This group of commands can help you correct your mistakes.
- They can be used in the current session only (i.e., once you close the file they will be useless).

Oops
- To retain the last group of erased objects.
- Works only with the **Erase** command.
- This command is only available in the Command window.
- You have to type the full command, meaning **oops**.
- No prompt will be dispalyed, except you will see the last group of erased objects coming back to the drawing.

Undo
- To undo the last command.
- You can reach this command using one of the following methods:
  - From the **Standard Annotation** toolbar, click the **Undo** button.
  - From menus select **Edit/Undo**.

- Type **u** at the Command window (don't type **undo**, because it has a different meaning).
- Press **Ctrl + Z**.
- The last command will be undone.
- You can undo as many commands as you want in the current session.

Redo

- To undo the undo.
- You can reach this command using one of the following methods:
  - From the **Standard Annotation** toolbar, click the **Redo** button.
  - From menus select **Edit/Redo**.
  - Type **redo** at the Command window.
  - Or press **Ctrl + Y**.
- The last undone command will be redone.
- You can redo as many commands as you want in the current session.

- To undo or redo a group of commands in one shot, do the following steps:
  - Select the pop-up list attached to the button.
  - The last command will be at the top of the list.
  - Move down to select the last command to be undone or redone.
  - Now a group of commands can be undone or redone in one shot.

REDRAW AND REGEN COMMANDS

- There will be times you will need to referesh the screen for one reason or another.
- Or you will need to make AutoCAD regenerate the whole drawing to show the arcs and circles as smooth curves.
- Both commands have no toolbar buttons.

Redraw
- From menus, select **View/Redraw** or type **r** at the Command window.
- The screen will be refreshed.

Regen
- To manually regenerate the drawing to show arcs and circles as smooth as they should be.
- From menus, select **View/Regen** or type **re** at the command prompt.
- See the example on the next page.

- This is how a drawing looks before the regen command:

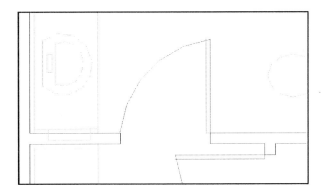

- And this is how it looks after the regen command:

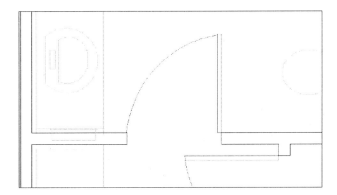

ERASE, UNDO, AND REDO

Exercise 11

1. Start AutoCAD 2008.
2. Open the file **Exercise_07.dwg**.
3. Using the **Erase** command, with Window or Crossing, do the following steps:
    a. Using Window, try to erase the two circles only. Use Undo to undo the erasing.
    b. Using Crossing, try to erase the two arcs only. Use Undo to undo the erasing.

c. Using pick box, try to erase the four lines. Use Undo to undo the erasing.
4. Close the file without saving.

CHAPTER REVIEW

1. Which of the following statements is true:
   a. Snap will help us control the mouse whereas Grid is complementary to Snap.
   b. ORTHO and Direct Distance Entry will help us draw exact orthogonal lines.
   c. We can use Direct Distance Entry with DYN, ORTHO, and POLAR.
   d. All of the above.
2. The Arc command in AutoCAD will draw a _____ arc.
3. In OTRACK, we can
   a. Specify a point using two existing points
   b. Specify the Radius of an arc
   c. Specify the end of an existing line
   d. None of the above
4. OTRACK doesn't need OSNAP to work:
   a. True
   b. False
5. In Polar, if the Increament angle didn't fulfill all your needs:
   a. ORTHO should help you
   b. Set the Additional angles
   c. The command POLARNEWANGLES will help
   d. None of the above.
6. There are _____ ways to draw a circle in AutoCAD.

CHAPTER REVIEW ANSWERS

1. d
2. Circular
3. a
4. b
5. b
6. 5 (five)

Chapter 3
HOW TO SET UP YOUR DRAWING

In This Chapter

- Things you need to know before setting up your drawing
- Setting up drawing units
- Setting up limits
- Creating and controlling layers

WHAT ARE THE THINGS TO THINK ABOUT?

- There are lots of things you need to think about while attempting to set up your drawing file. Of course we can't cover them all in this chapter, but we will mention the most important things.
- **Drawing Units** — As a first step, define the drawing distances and angle units, along with their precision.
- **Drawing Limits** — Try to figure out what size (area) of workspace will be sufficient to accommodate your drawing.
- **Layers** — Layers are the most effective way to organize your drawing, so we have to learn, what are they? How do we create them? How can we control them?
- In Appendix B we will discuss how to create the templates in AutoCAD, which are more applicable for establishments and not individuals.

STEP 1: DRAWING UNITS

- This is your first step.
- This command will allow you to select the proper length and angle units.

- From menus, select **Format/Units** or type **units**. The following dialog box will appear:

FIGURE 3-1

- Under **Length**, set up the desired **Type**. You will have 5 choices:
  - Architectural (example: 1'-5 3/16")
  - Decimal (example: 20.4708)
  - Engineering (example: 1'-4.9877")
  - Fractional (example: 17 1/16)
  - Scientific (example: 1.6531E+01)
- Under **Angle**, set up the desired **Type**. You will have 5 choices:
  - Decimal Degrees (example: 45.5)
  - Deg/Min/Sec (example: 45d30'30")
  - Grads (example: 50.6g)
  - Radians (example: 0.8r)
  - Surveyor's Units (example: N 45d30'30" E)
- For the desired Length and Angle select the **Precision**, for example:
  - Architectural precision can be 0'-0 1/16", or 0'-0 1/32", etc.
  - Decimal precision can be 0.00, or 0.000, etc.
  - Deg/Min/Sec precision can be 0d00'00", or 0d00'00.0", etc.
- By default AutoCAD deals with the positive angles as CounterClockWise. If you want it the other way around, switch on the checkbox **Clockwise**.

- Under **Insertion scale**, specify **Units to scale inserted content**, which is your drawing's scale against the scale of any object (a block for instance) in order for AutoCAD to make the suitable conversion.

- Click the **Direction** button to see the following dialog box:

FIGURE 3-2

- As we discussed in chapter 1, we said that AutoCAD always starts the zero angle measuring from the East. If you want to change it, select the desired angle to be considered as the new zero.

STEP 2: DRAWING LIMITS

- In Chapter 1, we said that AutoCAD offers us an unlimited drawing sheet, which extends in all directions. But we will not use it all; instead we will specify for ourselves an area, which will be our limits.
- Drawing limits is the workspace you select to work in and can be specified using two points—lower left corner and upper right corner.
- Since we will draw in Model Space and print from Paper Space we don't need to think about drawing scale at this point.
- To know exactly what the needed limits are, make sure you know the following information:
  - What is the longest dimension in your sketch in both X and Y.
  - What does AutoCAD unit mean to you (e.g., m, cm, mm, inch, or foot, etc.)
- Accordingly, we will know what the limit of your drawing is.

Example — Assume we have the following case:
- We want to draw an architectural plan, which extends in X for 50 m and in Y for 30 m.
- Also, assume that AutoCAD unit = 1 m.
- Since AutoCAD unit = 1 m, therefore, 50 m is equal to 50 AutoCAD units. The same thing applies for the 30 m.
- Note that 0,0 is always the favorite lower left corner, so no need to change it. The upper right corner will be 50,30.
- From menus, select **Format/Drawing Limits**, or type **limits**. The following prompt will appear at the Command window:

  ```
  Specify lower left corner or [ON/OFF] <0,0>:
  ```
  (press [Enter] to accept the default value)
  ```
  Specify upper right corner <12,9>:
  ```
  (type in the coordinate of the upper right corner)

ON/OFF — To forbid yourself from using any area outside this limit, turn the limits on.

STEP 3: LAYERS

What Are Layers?
- Let's assume that we have a huge number of transparent papers along with 256 colored pens.
- Taking care that we can't draw except on the paper at the top, we take the red pen and we draw a border.
- Then we move the second paper to the top and we draw an architectural wall plan using the yellow pen.
- Next we move the third paper to the top and we draw the doors using the green pen. Doing the same procedure we draw windows, furniture, electrical outlets, hatching, text, dimensioning, etc.
- Then we take all the papers and look at them at the same time, what you will see? A full architectural plan!
- In AutoCAD we call each paper a layer.
- Each layer should have a name, color, linetype, lineweight, and lots of other information.
- There is a layer in all of AutoCAD's drawings. This layer is 0 (zero). You can't delete it or rename it.
- In order to draw on a layer, make it **current** first. There will be only one layer as current.
- The objects drawn on a layer will automatically inherit the properties (color, linetype, and lineweight, etc.) of the current layer. Hence, a line

in a red layer, with dash-dot linetype, and 0.3 lineweight, will have the exact same properties.
- The setting of the object's color is by default = BYLAYER.
- The setting of the object's linetype is by default = BYLAYER.
- The setting of the object's lineweight is by default = BYLAYER.

- It is highly recommended to keep these settings intact, as changing them may lead you to create objects with nonstandard properties.

How to Create a New Layer

- You can start the layer command using one of the following methods:
  - From menus, select **Format/Layer**.
  - From the dashboard, and using the **Layers** panel, click **Layer Properties Manager**.
  - Type the **layer** command at the Command window.
- The following dialog box will appear:

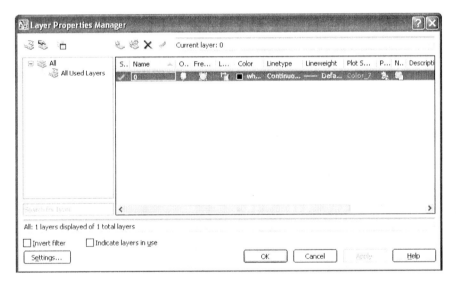

FIGURE 3-3

- Click the **New Layer** button or press Alt+N.
- AutoCAD will add a new layer with the temporary name *Layer1*. The **Name** field will be highlighted. Type the desired name of the layer (you can use up to 255 characters, spaces are allowed). Use ONLY the following:
  - Letters (a, b, c, . . . , z) small or capital doesn't matter.
  - Numbers (0, 1, 2, . . . , 9).
  - (-) hyphen, (_) underscore, and ($) dollar sign.

How to Set a Color for a Layer

- It is a common practice to use good layer naming, which gives an idea about the contents of the layer. As an example, a layer containing the walls of a building is named WALLS.
- After you create a layer, you set its color.
- AutoCAD uses 256 colors for the layers (as a matter of fact, they are only 255 if we exclude the color of the graphical area).
- The first seven colors can be called by their names or numbers:
  - Red (1)
  - Yellow (2)
  - Green (3)
  - Cyan (4)
  - Blue (5)
  - Magenta (6)
  - Black/White (7)
- The rest of the colors can be called only by their numbers.
- You can have the same color for more than one layer.
- Select the desired layer. Under the field **Color**, click either the name of the color or the icon. The following dialog box will appear:

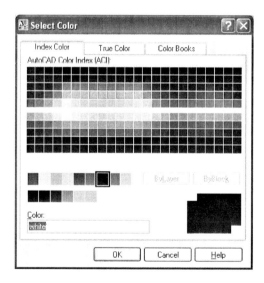

FIGURE 3-4

- Move to the desired color (or type in the name/number) of the color and then click **OK**.

How to Set a Linetype for a Layer

- AutoCAD comes with a good number of generic predefined linetypes saved in a file called *acad.lin*.
- You can also buy other linetypes from third-party vendors, who can be located on the Internet. Just go to any search engine (like Yahoo, Google, etc.) and search for "autocad linetype," and you will find lots of linetype files, some free of charge and some you can buy for few dollars.
- Not all of the linetypes are loaded in the drawing files; hence you may need to load the desired linetype first before you use it.
- Select the desired layer. Under the field **Linetype**, click the name of the linetype. The following dialog box will appear:

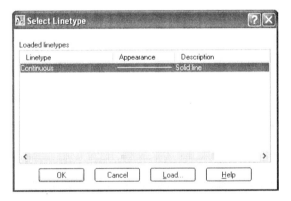

FIGURE 3-5

- By default only **Continuous** is loaded.
- To load another linetype, click the **Load** button. The following dialog box will appear:

FIGURE 3-6

- Select the desired linetype to be loaded and click **OK**. Now the linetype is loaded. It will appear next in the **Select Linetype** dialog box. Select it and click **OK**.

How to Set a Lineweight for a Layer
- Select the desired layer. Under the field **Lineweight**, click either the number or the shape of the lineweight. The following dialog box will appear:

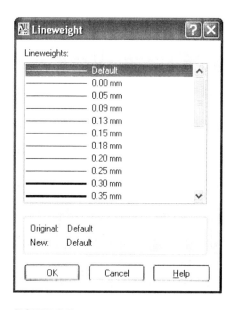

FIGURE 3-7

- Select the desired Lineweight and click **OK**.
- Using the **Status** bar, click the **LWT** button on if you want to view the lineweight of any layer on the screen.
- We prefer to see the lineweight using Plot Style (discussed in the last chapter), which will affect the hardcopy.

How to Make a Layer the Current Layer
- There are three ways to make a certain layer the current layer. They are:
  - In the **Layer Properties Manager** dialog box, double-click on the name or the status of the desired layer.
  - In the **Layer Properties Manager** dialog box, select the desired layer and click the **Set Current** button, or press Alt+C.

How to Set Up Your Drawing   61

- From the dashboard and using the **Layers** panel, use the layer's pop-up list, as shown below, to select the desired layer to be current:

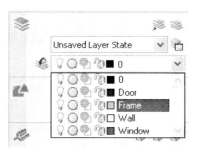

How to Add More Layers
- The easiest way is while you are in the **Layer Properties Manager**, the click on the name of any layer, then press [Enter].
- Or you can use the **New Layer** button.

- By default AutoCAD will always sort the layers according to their names.

How to Select Layers
- All of the methods discussed will be done in the **Layer Properties Manager** dialogue box.
- There are lots of ways to select layers:
  - To select a single layer, simply click on it.
  - To select multiple nonconsecutive layers, select the first layer, then hold the [Ctrl] key and click on the other layers.
  - To select multiple consecutive layers, select the first layer, then hold the [Shift] button and click on the last layer you wish to select.
  - To select multiple layers in one shot, click on an empty area and hold the mouse. Move it to the right or to the left and a rectangle will appear. Cover the layer that you wish to select and release the mouse.
  - To select all layers, press Ctrl+A.
  - To unselect a selected layer, hold the Ctrl key and click it.

- One of the most important advantages to selecting multiple layers is to set the color, linetype, or lineweight for a group of layers in one step.

How to Delete a Layer
- First, you have to know that you can't delete a layer that contains objects. So the first step is to empty any objects in the layer.
- Using the **Layer Properties Manager** dialog box, select the desired layer (or layers) to be deleted and do one of the following:
  - Press the [Del] key on the keyboard.

  - Click on the **Delete Layer** button or press Alt+D.

- Either way, AutoCAD will not delete these layers right away; instead, they will be marked for deletion. If you want to confirm the deletion do one of the following:
  - Click **OK**, the deletion will occur, and the dialog box will close.
  - Click the **Apply** button, the deletion will occur, but the dialog box will stay open.

How to Rename a Layer
- Select the layer you want to rename.
- Click (one click only) on the name—the name will be available for editing. Type in the new name and press [Enter].

What Will Happen if You Right-click?
- Right-clicking here is done in the **Layer Properties Manager** dialog box.
- If you select any layer and right-click, the following shortcut menu will appear:

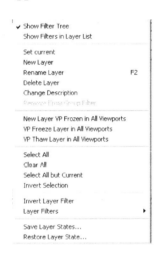

- Through this shortcut menu you can do lots of things we discussed earlier such as:
  - Set the current layer.
  - Create a new layer.
  - Delete a layer.
  - Select All layers.
  - Clear the selection.
  - Select All but Current.
  - Invert Selection (make the selected unselected and vice versa).
- The first two choices in this shortcut menu are:
  - Show Filter Tree (by default is turned on)
  - Show Filters in Layer List (by default is turned off)

- By turning off **Show Filter Tree**, the dialog box will have more space, just like the following:

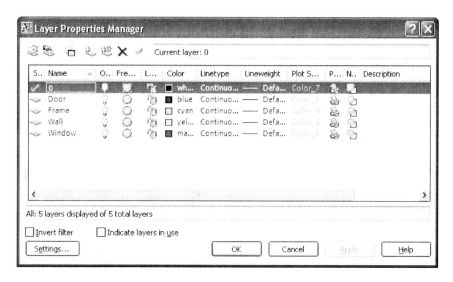

FIGURE 3-8

- You can resize this dialog box by going to any of the four edges. Once the mouse pointer changes, click and hold and make the dialog box larger or smaller.

How to Change Object's Layer

- Each object should exist in a layer.
- The fastest method to change the object's layer is the following:
  - Without issuing any command, select the object (by clicking it).
  - In the **Layers** toolbar, the object's layer is displayed. To change it, click the layer's pop-up list and select the new layer.
  - Press [Esc] one time.
- There are other methods that will be discussed later.

How to Make an Object's Layer Current

- This function is very handy in case there are too many layers in your drawing.
- You see an object in your drawing, but you don't know in which layer this object exists!
- What you want is to make this object's layer the current layer. To do that follow these steps:
  - From the dashboard and using the **Layers** panel, click the **Make Object's Layer Current** button.

What Are the Four Switches of a Layer?

- The following prompt will be shown:

  ```
  Select object whose layer will become current:
  (click on the desired object)
  Walls is now the current layer.
  ```

- Now the current layer is the object's layer.
- Each layer has four switches, which will determine its state.
- You can see these switches in both the **Layer Properties Manager** dialog box and the layer pop-up list from the Layer toolbar.
- These switches are:
  - On/Off switch.
  - Thaw/Freeze switch.
  - Unlock/Lock switch.
  - Plot/No Plot switch.
- These four switches are independent of each other.
- By default, the layers are On, Thaw, Unlock, and Plot.
- When you turn a layer off, then the objects in it will not be shown on the screen, and if you plot the drawing, they will not be plotted. But the objects in this layer will be counted in the total count of the drawing; hence the drawing size will not change.
- When you freeze a layer, then the objects in it will not be shown on the screen, and if you plot the drawing, they will not be plotted. Also, the objects in this layer will NOT be counted in the total count of the drawing; hence the drawing size will be less.
- When you lock a layer, then none of the objects in it are modifiable.
- When you make a layer no plot that means you can see the objects on the screen, but when you issue a **Plot** command these objects will not be plotted.
- Three of these switches can be changed using both the **Layer Properties Manager** dialog box, and layer pop-up list from the **Layer** toolbar; the fourth (i.e., Plot/No Plot) can be changed only from the **Layer Properties Manager** dialog box.
- To change the switch, simply click it.
- You can't freeze the current layer. The following dialog box will appear:

FIGURE 3-9

- But you can turn off the current layer with a warning message:

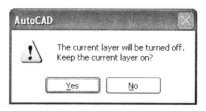

FIGURE 3-10

- Here AutoCAD will assume you did this by mistake, so it will ask you to keep the current layer on. If you click **Yes**, nothing will happen. On the other hand, if you click **No**, the current layer will be turned off.
- You should be careful when you turn the current layer off, because each and every time you draw a new object it will disappear right away.

What is Layer Previous?

- While you are working in AutoCAD, you will change the state of layers a lot, which means you need a tool to help you retain the previous state.
- **Layer Previous** is the answer for such a requirement.

- From the dashboard and using the **Layers** panel, click the **Layer Previous** button.
- AutoCAD will report the following statement:

```
Restored previous layer status.
```

- While you are in the Layer Properties Manager dialog box or layer pop-up list from the Layer toolbar, and you make several changes on several switches for several layers, AutoCAD will consider them one action; hence it will restore them in one Layer Previous command.

CHANGING AN OBJECT'S PROPERTIES AND MATCH PROPERTIES

- Earlier in this chapter, we said that each object will inherit the properties of the layer in which it resides. By default the settings of the current Color, Linetype, and Lineweight is BYLAYER, which means that the object follows the layer it exists in.
- This makes control of the drawing easier, as it is easier to control a handful of layers rather than controlling hundreds of thousands of objects.
- So we recommend to not change these settings and to stick with them in normal times.
- Knowing that certain demands may force us to break some rules, we may use the Object's **Properties** palette, which will help us make some changes to the object(s).

Properties ▪ The easiest way to initiate this command is to select the desired object(s) and then right-click. When the shortcut menu appears, select **Properties**. There are two possibilities:
  • The selection set you made consists of different object types (lines, arcs, circles, etc.). In this case you can change only the **General** properties of these objects. The following will appear:

FIGURE 3-11

  • The selection set you made consists of a single object type. In this case you can change the **General** properties and object-sepcific properties. The following will appear:

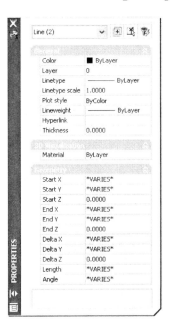

FIGURE 3-12

How to Set Up Your Drawing 67

Other Ways to Issue the Properties Command

- There are four more ways to issue the command **Properties**.
  - From the **Standard Annotation** toolbar, click the **Properties** button.
  - From menus select **Tools/Palettes/Properties**.
  - Press **Ctrl+1** (number 1 at the top and not at the right side).
  - Type **properties** on the keyboard.

Match Properties

- If you open a drawing and find that the creator of this drawing didn't stick with the concept of BYLAYER—for example you find a green line resides in a red layer, and a dash-dot circle is in a layer with continuous linetype—the best way to correct such misconduct is to try to find one object in each layer that holds the right properties and then match the other objects to it.
- To issue this command, use one of the following:

  - From the **Standard** toolbar, click the **Match Properties** button.
  - From menus, select **Modify/Match Properties**.
  - Type **matchprop** on the keyboard.
- Either way, AutoCAD will show the following prompt:

```
Select source object:
```

- Click on the object that holds the right properties.
- The mouse pointer will change to a brush shape and AutoCAD will prompt:

```
Select destination object(s):
```

- Click on the objects you want to correct. Once you are done, press [Enter].

USING LAYERS

Exercise 12

1. Start AutoCAD 2008.
2. Open the file **Exercise_12.dwg**.
3. Create the following four layers:

Layer Name	Color	Linetype
Shaft	Magenta	Continuous
Body	Cyan	Continuous
Base	Green	Continuous
Centerlines	Yellow	Dashdot2

4. Make **Centerlines** current (make sure that **DYN** is off).
5. Draw a line from 6,7.5 to 6,4.5, then draw another line from 8,6 to 4,6.
6. Using any of the methods you learned, change the object's layer as follows:
   a. Change the layer of the two circles from 0 to Shaft.
   b. Change the layer of the two arcs from 0 to Body.
   c. Change the layer of the lines from 0 to Base.
7. Lock layer Shaft. Then try to erase the objects in it. What is the response of AutoCAD?
8. Unlock layer Shaft.
9. Try to freeze the current layer (it doesn't matter which one!). What is the response of AutoCAD?
10. Try to rename layer 0? What is the response of AutoCAD?
11. Rename layer **Centerlines** to be **Center_lines**.
12. Try to delete layer Shaft. What is the response of AutoCAD? Type it in: _____, and Why?
13. Save the file and close it.

CREATING OUR PROJECT (METRIC)

Workshop 1-A

1. Start AutoCAD 2008.
2. Close any opened file.
3. Create a new file based on the *acad.dwt* template.
4. Double-click on the mouse wheel to zoom extents.
5. Select **Format/Units** and make the following changes:
   a. Length Type = Decimal, Precision = 0
   b. Angle Type = Decimal Degrees, Precision = 0
   c. Unit to scale inserted content = Millimeters
6. Assume that AutoCAD unit = 1 mm, and you have a 30 x 20 m plan you want to draw, therefore your limits will be:
   d. Lower left corner = 0,0
   e. Upper right corner = 30000,20000
7. Select **Format/Drawing Limits** and set the limits accordingly.
8. Double-click on the mouse wheel to zoom extents.
9. Create the following layers:

Layer Name	Color	Linetype	Special Remarks
Frame	Magenta	Continuous	
Walls	Red	Continuous	
Doors	Yellow	Continuous	
Door_Swing	Yellow	Dashed	
Windows	150	Continuous	
Centerlines	Green	Dashdot	
Bubbles	Green	Continuous	
Furniture	41	Continuous	
Staircase	140	Continuous	
Text	Cyan	Continuous	
Hatch	White	Continuous	
Dimension	Blue	Continuous	
Viewports	8	Continuous	No Plot

10. Save the file as **Small_Villa_Ground_Floor_Plan_Metric.dwg**.

CREATING OUR PROJECT (IMPERIAL)

Workshop 1-B

1. Start AutoCAD 2008.
2. Close any opened file.
3. Create a new file based on the *acad.dwt* template.
4. Double-click on the mouse wheel to zoom extents.
5. Select **Format/Units** and make the following changes:
    a. Length Type = Architectural, Precision = 0'-0"
    b. Angle Type = Decimal Degrees, Precision = 0
    c. Unit to scale inserted content = Inches
6. Assume that AutoCAD unit = 1 inch, and you have 70' x 60' plan you want to draw, therefore your limits will be:
    d. Lower left corner = 0,0
    e. Upper right corner = 70',60'
7. Select **Format/Drawing Limits** and set the limits accordingly.
8. Double-click on the mouse wheel to zoom extents.

9. Create the following layers:

Layer Name	Color	Linetype	Special Remarks
Frame	Magenta	Continuous	
Walls	Red	Continuous	
Doors	Yellow	Continuous	
Door_Swing	Yellow	Dashed	
Windows	150	Continuous	
Centerlines	Green	Dashdot	
Bubbles	Green	Continuous	
Furniture	41	Continuous	
Staircase	140	Continuous	
Text	Cyan	Continuous	
Hatch	White	Continuous	
Dimension	Blue	Continuous	
Viewports	8	Continuous	No Plot

10. Save the file as **Small_Villa_Ground_Floor_Plan_Imperial.dwg**.

CHAPTER REVIEW

1. What is true about layer names:
    a. They can be up to 255 characters.
    b. They can use spaces in their name.
    c. They can use letters, numbers, hyphens, underscores, and dollar signs.
    d. All of the above.
2. There are _____ different Length units in AutoCAD.
3. What do you need to know to set up your limits in a certain file:
    a. What is the paper size you will print on?
    b. What is the longest dimension of your sketch in both X and Y.
    c. AutoCAD unit equals to what?
    d. B & C.
4. Only the first 7 colors can be called by their name and number:
    a. True
    b. False

5. What is true about linetypes in AutoCAD?
   a. They are stored in acad.lin.
   b. They are loaded in all AutoCAD drawings.
   c. If you need to use a linetype you have to load it first.
   d. A & C.
6. If you assigned a lineweight to a layer and on this layer you draw lines, you need to switch _____ from status bar to see this lineweight on the screen.
7. You can change only the _____ properties of nonsimilar objects using the Properties command.

CHAPTER REVIEW ANSWERS

1. d
2. 5 (five)
3. d
4. a
5. d
6. LWT
7. General

Chapter 4
A Few Good Construction Commands

In This Chapter

- Creating a parallel duplicate using the Offset command
- Creating neat intersections using the Fillet and Chamfer commands
- Trimming and Extending objects
- Lengthening objects

INTRODUCTION

- Up until now we learned only 4 drawing commands (namely; line, arc, circle, and polyline).
- These alone can't help you accomplish merely 20% of your drawing.
- Also, if you think that each and every line (or arc, or circle) should be drawn by you, you are wrong!
- In this chapter, we will discuss six commands that will magically help us construct the most difficult drawings in no time.
- These commands are:
  - Offset command, creates parallel copies of your original objects.
  - Fillet command, allows you to close unclosed shapes either by extending the two ends to an intersecting point or by arc.
  - Chamfer command, exactly the same as Fillet except it will create a beveled edge (an edge with slope).
  - Trim command, allows some objects to act as cutting edges for other objects to be trimmed.
  - Extend command, allows you to extend objects to a boundary.
  - Lengthen command, allows you to extend or trim length from an existing line.

OFFSET COMMAND

- Offset will create a new object parallel to the selected object.
- The new object (by default) will have the same properties of the original object and will reside in the same layer.
- There are two methods to offset:
  - Using offset distance.
  - Using a Through point.
- To issue this command, use one of the following methods:
  - From the dashboard and using **2D Draw** panel, click the **Offset** button.
  - From menus select **Modify/Offset**.
  - Type **offset** (or **o**) in the Command window.
- The following prompt will appear:

```
Specify offset distance or [Through/Erase/Layer]
<Through>:
```

Offset Distance
- If you want to use this method, you should know the distance between the original object and the parallel duplicate (i.e., offset distance).
- Then select object, which will be offset.
- Finally, specify the side of offset by clicking (to the right, or left, up, or down, etc.)
- The prompts will be as follows:

```
Specify offset distance or [Through/Erase/Layer]
<Through>: (type in the desired distance)
Select object to offset or [Exit/Undo] <Exit>:
(select a single object)
Specify point on side to offset or [Exit/Multiple/
Undo] <Exit>: (click in the desired side)
```

- Command will repeat the last two prompts for further offsetting.
- To end the command press [Enter] or right-click.
- Here is an example:

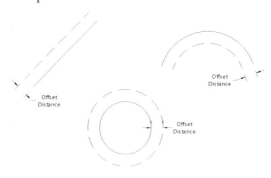

Through Point
- With this method, there is no need to know the distance but rather to know any point that the new parallel image will pass through.
- The prompt will be as follows:

  ```
  Specify offset distance or [Through/Erase/Layer]
  <Through>: (type t)
  Select object to offset or [Exit/Undo] <Exit>:
  (Select a single object)
  Specify through point or [Exit/Multiple/Undo]
  <Exit>: (Specify the point that the new image will
  pass through)
  ```

- Command will repeat the last two prompts for further offsetting.
- To end the command press [Enter] or right-click.
- Here is an example:

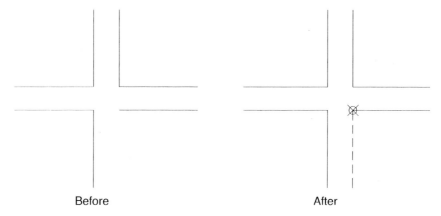

Before   After

Multiple
- With both methods discussed above you can use the **Multiple** option.
- If you have the following case:

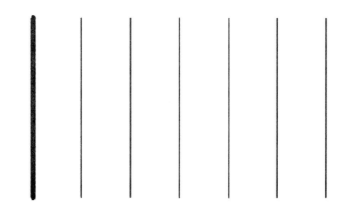

- The original object is the thick one and the rest are offset objects. As you can see, the distance between all objects is the same, hence instead of having to keep selecting the object and specifying the side, the **Multiple** option will allow you only to specify the side of offset.
- The prompt will be as follows:

  ```
  Specify offset distance or [Through/Erase/Layer]
  <Through>: (Select either method)
  Select object to offset or [Exit/Undo] <Exit>:
  (Select a single object)
  Specify through point or [Exit/Multiple/Undo]
  <Exit>: (Type m)
  Specify point on side to offset or [Exit/Undo]
  <next object>: (simply click on the desired side
  and you can keep doing the same, once you are done
  press [Enter])
  ```

- At any moment while you are offsetting you can use the **Undo** option to undo the last offsetting action.

- AutoCAD will recall the last Offset distance, so no need to keep typing it unless you want to use another value.
- Offset will produce either a bigger or smaller arc, circle, or polyline.
- You can right-click to show shortcut menus displaying the different options of the Command window.
- In the same Offset command you can use only one offset distance. If you want another offset distance, end the current command and issue a new offset command. (*We hope to see in the next version of AutoCAD an Offset command that will allow the user to use more than one offset distance per command.*)

OFFSETTING OBJECTS

Exercise 13

1. Start AutoCAD 2008.
2. Open the file **Exercise_13.dwg**.
3. Offset the walls (magenta) to the inside using the distance = **1'**.
4. Offset the stairs using distance = **1'–6"** and using the **Multiple** option to create **8** lines representing 8 steps.
5. Explode the inner polyline.

6. Offset the right vertical line to the left using the **Through** option and the left end point of the upper right horizontal line.
7. Offset the new line to the right using distance = **6"**.
8. The new shape of the plan should look like the following:

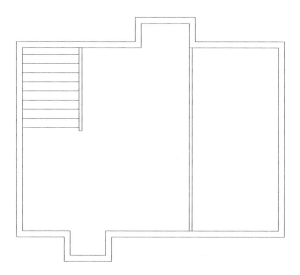

FILLET COMMAND

- If you have the following cases:

- And you want them to look like this:

- Or you want them to look like this:

- Then you definitely need to use the **Fillet** command.
- Issue the **Fillet** command, select the first object and then the second object. It is a very simple AutoCAD command.
- The Fillet command works in two different settings:
  - Radius = 0 will create a neat intersection.
  - Radius > 0 will do as above except using an arc.
- When you want to close the shape with an arc, what will happen to the original objects? To solve this issue, Fillet command works in two different modes:
  - **Trim**, the arc will be produced, and the original objects will be trimmed accordingly.
  - **No trim**, the arc will be produced, but the original objects will stay intact.
- Here is an example:

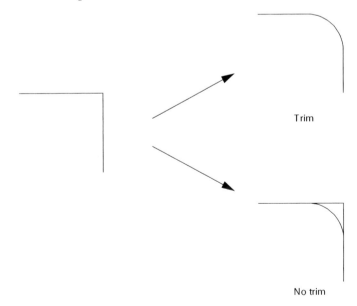

Trim

No trim

A Few Good Construction Commands 79

- Use one of the following to start the Fillet command:
  - From the dashboard and using **2D Draw** panel, click on the small triangle on the right, keep holding, then click the **Fillet** button.
  - From menus select **Modify/Fillet**.
  - Type **fillet** (or **f**) in the Command window.
- The following prompt will appear:

```
Current settings: Mode = TRIM, Radius = 0.0000
Select first object or [Undo/Polyline/Radius/Trim/
Multiple]:
```

- The first line reports the current value of Mode and Radius.
- Choose between the different options:

Radius
- To set a new value for the Radius the following prompt will appear:

```
Specify fillet radius <0.0000>:
```
(type in the new radius)

Trim
- To change the mode from Trim to No trim and vice versa, the following prompt will appear:

```
Enter Trim mode option [Trim/No trim] <Trim>:
```
(type t, or n)

Multiple
- By default you can perform a single fillet per command by selecting the first object and the second object.
- If you want to perform multiple fillets in a single command you have to select this mode first.

Undo
- At any moment you are filleting you can use the **Undo** option to undo the last filleting action.

NOTE
- When you fillet with a radius, the radius will be created in the current layer. So make sure that you are in the right layer.
- To end the command when you use the **Multiple** option, press [Enter] or right-click.
- Even if R > 0 you can still fillet with R = 0. To do that, simply hold [Shift] key and click on the objects desired, regardless of the current value of R you will fillet with R = 0.
- You can use the Fillet command to fillet two parallel lines with an arc. AutoCAD will calculate the distance between the two lines and take the Radius to be one-half of this length.

FILLETING OBJECTS

Exercise 14

1. Start AutoCAD 2008.
2. Open the file **Exercise_14.dwg**.
3. Make sure that the current layer is *Base*.
4. Using the Fillet command, make the following steps:
   a. Set the radius = 0.5
   b. Mode = Trim
5. Now fillet the right vertical line with the righthand arc and the left vertical line with the lefthand arc (the result may not be perfect but we will fix it later). You should get something like the following:

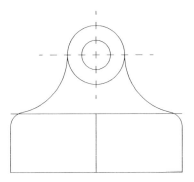

6. Using the Fillet command, make the following steps:
   a. Set the radius = 0.4
   b. Mode = No Trim
   c. Set Fillet to be multiple
7. Now fillet the inner vertical line with the two horizontal lines to get the following result:

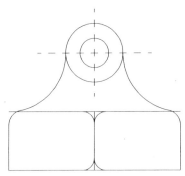

8. Save the file and close it.

CHAMFER COMMAND

- Identical in many things to Fillet, except it creates a sloped edge rather than creating an arc.
- To create the sloped edge, we will use one of two methods:
  - Two distances.
  - Length and Angle.

Two Distances
- There are 3 different cases for this method:
  - (Dist1 = Dist2) = 0.0, as in the following case:

  - (Dist1 = Dist2) > 0.0, as in the following case:

  - (Dist1 ≠ Dist2) > 0.0, as in the following case (for whichever is selected first, Dist1 is used):

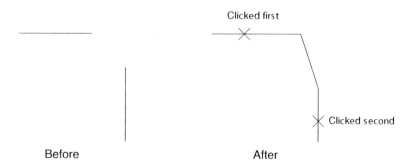

Length and Angle — To use this method, specify a length (which will be removed from the first object) and an angle, just like the following example:

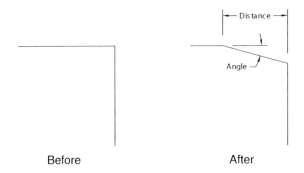

Before | After

- Use one of the following to start the Chamfer command:
  - From the dashboard and using **2D Draw** panel, click the small triangle on the right, keep holding, then click the **Chamfer** button.
  - From menus select **Modify/Chamfer**.
  - Type **chamfer** (or **cha**) in the Command window.
- The following prompt will appear:

```
(TRIM mode) Current chamfer Dist1 = 0.0000, Dist2 = 0.0000
Select first line or [Undo/Polyline/Distance/Angle/Trim/mEthod/Multiple]:
```

- The first line reports the current Mode and Distances (or Length and Angle).
- Choose between the different options:

Distances — To set a new value for the **Distances**, the following prompt will appear:

```
Specify first chamfer distance <0.0000>: (input the first distance)
Specify second chamfer distance <0.0000>: (input the second distance)
```

Angle — To set the new values for both length and angle, the following prompt will appear:

```
Specify chamfer length on the first line <0.0000>: (input the length on first line)
Specify chamfer angle from the first line <0>: (input the angle)
```

Trim — To change the mode from Trim to No trim and vice versa, the following prompt will appear:

A FEW GOOD CONSTRUCTION COMMANDS   83

                Enter Trim mode option [Trim/No trim] <Trim>: *(type t, or n)*

Method ▪ To specify the default method used in the Chamfer command, the following prompt will appear:

                Enter trim method [Distance/Angle] <Distance>: *(type d, or a)*

Multiple ▪ By default you can perform a single chamfer per command, by selecting the first object and the second object. If you want to perform multiple chamfers in a single command you have to select this mode first.

 ▪ When you chamfer, the sloped line will be created in the current layer. So make sure that you are in the right layer.

      ▪ To end the command when you use the **Multiple** option, press [Enter] or right-click.

CHAMFERING OBJECTS

Exercise 15

1. Start AutoCAD 2008.
2. Open the file **Exercise_15.dwg**. (If you solved the previous exercise correctly, open your file.)
3. Make sure that the current layer is *Base*.
4. Using the Chamfer command, make the following steps:
   a. Set Dist1 = 1.0
   b. Set Dist2 = 0.4
   c. Mode = Trim
   d. Set Chamfer to be multiple
5. Now chamfer the lower edges by selecting the horizontal line first, then the vertical line, to produce the following shape:

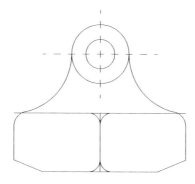

6. Save the file close it.

TRIM COMMAND

- Trimming means we want to remove part of an object according to a cutting edge(s).
- The Trim command is a two-step command:
  - The first step is to select the cutting edge(s). It can be one object, or as many as you wish.
  - The second step is to select the objects to be trimmed.
- The following example will illustrate the process of trimming:

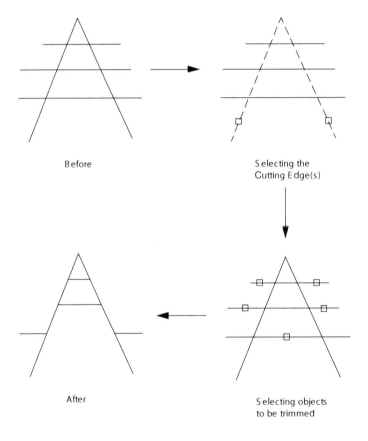

- Use one of the following to start the Trim command:
  - From the dashboard and using **2D Draw** panel, click on the small triangle at the right, keep holding, then click the **Trim** button.
  - From menus select **Modify/Trim**.
  - Type **trim** (or **tr**) in the Command window.

- The following prompt will appear:

  ```
  Select cutting edges ...
  Select objects or <select all>:
  ```

- The first line is telling you to select the cutting edges.
- Use any of the methods we learned in the Erase command. Once you are done press [Enter] or right-click.
- You can also use the fastest way, the **select all** option, which will select all the objects.
- The following prompt will appear:

  ```
  Select object to trim or shift-select to extend or
  [Fence/Crossing/Project/Edge/eRase/Undo]:
  ```

- Now click on the parts you want to trim, one-by-one.
- If you made any mistakes, either right-click to bring up the shortcut menu and select **Undo**, or type **u**.

Fence
- You can use the **Fence** option to speed up your process of selecting the objects to be trimmed. This can be done by specifying 2 points or more, creating a dotted line. Whatever objects it touches will be trimmed.

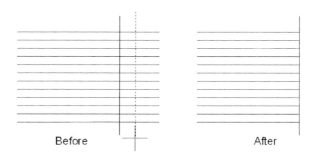
Before   After

Crossing
- The same thing applies for the **Crossing** option, but this can be done by specifying two opposite corners. A crossing window will appear and any object touched by the crossing will be trimmed.

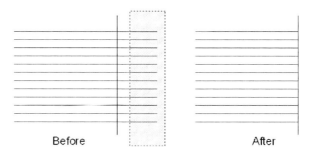
Before   After

Erase
- Sometimes, as a result of trimming, some objects will be unwanted. So, instead of finishing the command and issuing an **Erase** command, AutoCAD made this option availabe for you to erase objects while you are still in Trim command.
- Type **r**, and AutoCAD will ask you to select the objects you want to erase. Once you are done, press [Enter]. The prompts will appear again to select another option.

TRIMMING OBJECTS

 Exercise 16

1. Start AutoCAD 2008.
2. Open the file **Exercise_16.dwg**. (If you solved the previous exercise correctly, open your file.)
3. Using the Trim command, try to get the following shape:

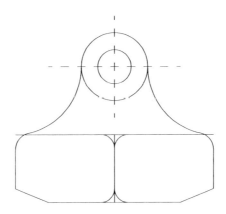

4. Save the file and close it.

EXTEND COMMAND

- It is the opposite of the **Trim** command.
- You will extend selected objects to boundary edge(s).
- The Extend command is a two-step command:
  - The first step is to select the boundary edge(s). It can be one object, or as many as you wish.
  - The second step is to select the objects to be extended.

- The following example will illustrate the process of extending:

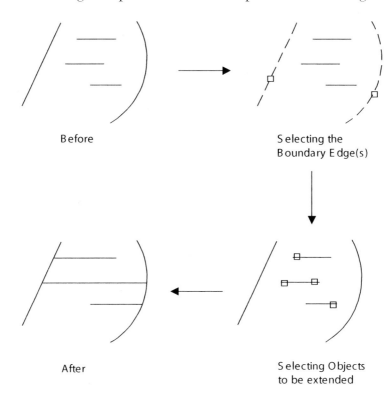

Before

Selecting the Boundary Edge(s)

After

Selecting Objects to be extended

- Use one of the following to start the Trim command:
  - From the dashboard and using **2D Draw** panel, click on the small triangle at the right, keep holding, then click the **Extend** button.
  - From menus select **Modify/Extend**.
  - Type **extend** (or **ex**) in the Command window.
- The following prompt will appear:

```
Select boundary edges ...
Select objects or <select all>:
```

- The first line is telling you to select the boundary edges.
- Use any of the methods you know. Once you are done press [Enter] or right-click.
- The following prompt will appear:

```
Select object to extend or shift-select to trim
or [Fence/Crossing/Project/Edge/Undo]:
```

- Now click on the parts you want to extend, one-by-one.

- If you made any mistakes, either right-click to bring up the shortcut menu and select **Undo**, or type **u**.
- The rest of the options are just like Trim command.
- **NOTE** While you are in the **Trim** command, and while you are clicking on the objects to be trimmed, if you hold the [Shift] key and click, you will be extending the objects and not trimming them, and vice versa.
- See the below example:
  - Assume you have the following case:

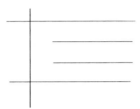

  - Start the Trim command. Select the vertical line as the cutting edge, then press [Enter].
  - Click the left part of both the upper and lower horizontal lines.
  - The situation will be as follows:

  - We are still in the Trim command. Hold the [Shift] key and click the two intermediate horizontal lines, they will extend even though you are in the Trim command. You will have the following picture:

EXTENDING OBJECTS

Exercise 17

1. Start AutoCAD 2008.
2. Open the file **Exercise_17.dwg**.
3. Using the Extend and Trim commands, try to get the following shape:

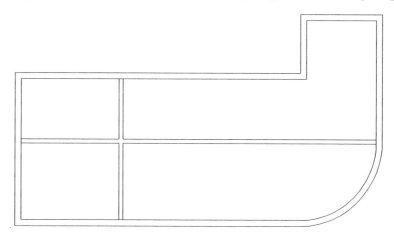

4. Save the file and close it.

LENGTHEN COMMAND

- In the Extend command, we needed an object to serve as a boundary in order to extend the rest of the objects to it.
- Lengthen (or shorten, as it serves both purposes) will do this without the need of any boundary.
- Use one of the following methods to issue the command:
  - From menus select **Modify/Lengthen**.
  - Type **lengthen** (or **le**) in the Command window.
- Either way, the following prompt will appear:

  ```
  Select an object or [DElta/Percent/Total/DYnamic]:
  ```

- If you click on any object, it will give you the current length.
- The command will do the lengthening (or shortening) using the following methods:

Delta
- You want to add (remove) to the current length an extra length.
- If you input a negative value, the Lengthen command will shorten the line.

- The following prompt will appear:

  Enter delta length or [Angle] <0.0000>: *(input the extra length to be added)*

Percent
- You want to add (remove) from the length by specifying a percentage of the current length.
- The number should be positive and a nonzero number. If it is > 100, it will lengthen. If it is < 100, it will shorten.
- The following prompt will appear:

  Enter percentage length <100.0000>: *(input the new percentage)*

Total
- You want the new total length of the line to be equal to the number you input.
- If the new number > current length, the line will lengthen. If the new number < current length, the line will shorten.
- The following prompt will appear:

  Specify total length or [Angle] <1.0000)>: *(input the new total length)*

Dynamic
- To specify a new length of the object using a dynamic move of the mouse.
- The following prompt will appear:

  Select an object to change or [Undo]: *(select the desired object)*

  Specify new end point: *(move the mouse, up until you reach to the desired length)*

- You can use only one method per command.

LENGTHENING OBJECTS

Exercise 18

1. Start AutoCAD 2008.
2. Open the file **Exercise_18.dwg**.
3. Using the Lengthen command and **Delta** option, shorten the two vertical lines by **1** unit.
4. Using the Lengthen command and **Total** option, make the two horizontal lines total length = 5.
5. As you can see, the lower line didn't come to the end like the upper line.
6. Using the Lengthen command and **Percent** option, using Percent = **104**, select the end of the line.

7. The output should look like the following:

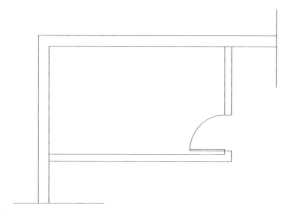

DRAWING THE PLAN (METRIC)

 Workshop 2-A

1. Start AutoCAD 2008, and open the file: **Small_Villa_Ground_Floor_Plan_Metric_Workshop_2.dwg**. (If you solved the previous workshop correctly, open your file.)
2. Make the **Walls** layer current.
3. Using the Polyline command, draw the lines first (without the dimensions) starting from point 8000,3000 using all the methods you learned in Lesson 2.

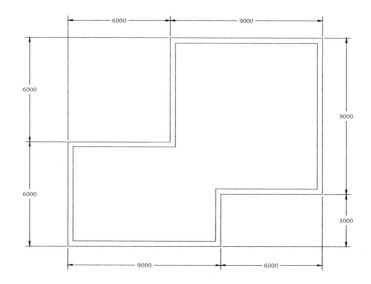

4. Using the Offset command, offset the polyline to the *inside* with offset distance = **300**.
5. Explode the inner polyline.
6. Using the Offset, Fillet, Chamfer, Trim, Extend, Lengthen, and Zoom commands try to make the interior walls using the following dimensions:

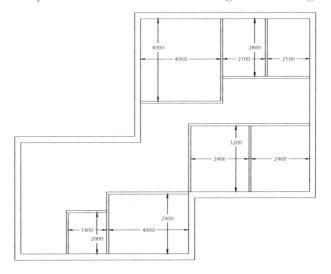

 ▪ The thickness of all inner walls is = 100.

7. Make the door openings as follows, taking into consideration the following:
   a. All door openings are 900.
   b. Always take 100 clear distance from the walls for the door openings, except for the outside door take 500.

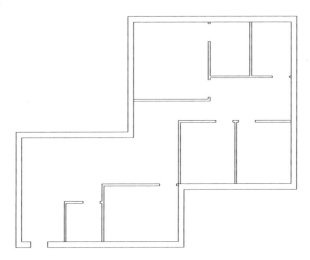

■ To make the door openings, use the following technique:
- Offset an existing wall (say 100 for internal doors).
- Offset the new line (say 900 for room doors).
- You will have the following shape:

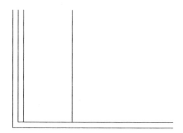

- Extend the two vertical lines to the lower horizontal line, just like this:

- Using the Trim command, select all the horizontal lines and vertical lines as cutting edges, then press [Enter].
- As for the objects to trim, click the following parts (you can use Crossing, which is more speedy).

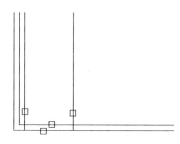

- This is what you will get:

8. Save the file and close it.

DRAWING THE PLAN (IMPERIAL)

 Workshop 2-B

1. Start AutoCAD 2008, and open the file: **Small_Villa_Ground_Floor_Plan_Imperial_Workshop_2.dwg**. (If you solved the previous workshop correctly, open your file.)
2. Make the **Walls** layer current.
3. Using the Polyline command, draw the lines (without the dimensions) starting from point 16',10' using all the methods you learned in Lesson 2.

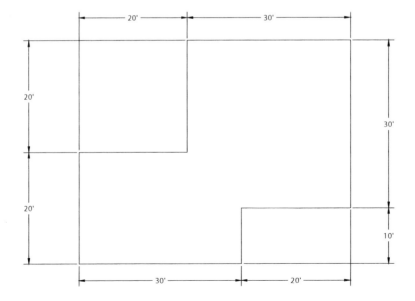

4. Using the Offset command, offset the polyline to the *inside* with offset distance = **1'**.

5. Explode the inner polyline.
6. Using the Offset, Fillet, Chamfer, Trim, Extend, Lengthen, and Zoom commands try to make the interior walls using the following dimensions:

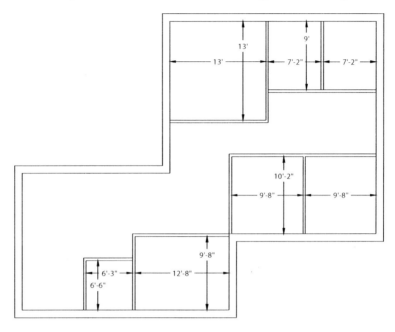

 The thickness of all inner walls is = 4".

7. Make the door openings as follows taking into consideration the following:
   a. All door openings are 3'.
   b. Always take 4" clear distance from the walls for the door openings, except for the outside door take 1'6".

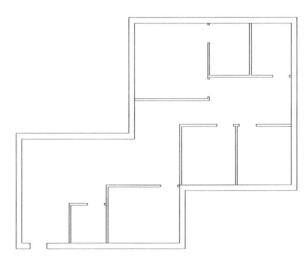

■ To make the door openings, use the following technique:
* Offset an existing wall (say 4" for internal doors).
* Offset the new line (say 3' for room door).
* You will have the following shape:

* Extend the two vertical lines to the lower horizontal line, just like this:

* Using the Trim command, select all the horizontal lines and vertical lines as cutting edges, then press [Enter].
* As for the objects to trim, click the following parts (you can use Crossing, which is more speedy).

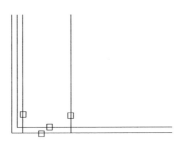

- This is what you will get:

8. Save the file and close it.

CHAPTER REVIEW

1. Using the same Offset command you can use more than one offset distance:
   a. True, only two offset distances.
   b. True, you can use as many as you wish.
   c. False, only one offset distance.
   d. The only method available in Offset command is the Through point method.
2. In the Lengthen command, using the Percent option, 150% should be input as _____.
3. While trimming you can extend and vice versa:
   a. True
   b. False
4. You can fillet using an arc, but you need to specify:
   a. Distances
   b. Radius
   c. Radius and Distances
   d. Length and Angle
5. There are two methods to Chamfer—Distances and Length/Angle:
   a. True
   b. False
6. The first step in the Extend command is to select _____, and the second step is to select _____.

CHAPTER REVIEW ANSWERS

1. c
2. 150
3. a
4. b
5. a
6. Boundary Edge(s), Objects to extend

Chapter 5
MODIFYING COMMANDS

In This Chapter

- Advanced techniques in selecting objects
- Moving and copying objects
- Rotating and scaling objects
- Creating duplicates using the Array command
- Mirroring objects
- Stretching objects
- Breaking objects
- Using Grips to modify objects

INTRODUCTION

- In this chapter we will learn the core of the Modifying commands in AutoCAD.
- We will cover nine commands, which will enable the user to make any type of changes on the drawing.
- First we will discuss the selection process (more advanced than what we discussed in Chapter 2).
- Then we will discuss the following commands:
  - Move command, to move objects from one place to another.
  - Copy command, to copy objects.
  - Rotate command, to rotate objects using rotation angles.
  - Scale command, to create bigger or smaller objects using a scale factor.
  - Array command, to create copies of objects either in a matrix fashion, or circular or semicircular fashion.
  - Mirror command, to create mirror images of selected objects.
  - Stretch command, to change the length of objects.
  - Break command, to break an object into two pieces.
- We will wrap up this subject discussing the Grips in AutoCAD.

SELECTING OBJECTS

- All of the Modifying commands (with some exceptions) will ask you the same question:

  ```
  Select objects:
  ```

- In Chapter 2 we took a look at some of the methods to select. What we will do now is expand our knowledge in this area.
- With all the methods we will discuss we will type a letter or more at the **Select objects** prompt.

W (Window)
- If you typed **W**, the mode **Window** will be available whether you went to the right or to the left.

C (Crossing)
- If you typed **C**, the mode **Crossing** will be available whether you went to the right or to the left.

WP (Window Polygon)
- Excellent choice when you are drawing using other angles like 30, 45, 60, etc.
- When you type **WP** and press [Enter], the following prompt will appear:

  ```
  First polygon point:
  ```
  (specify the first point of the polygon)
  ```
  Specify endpoint of line or [Undo]:
  ```
  (specifying the second point)
  ```
  Specify endpoint of line or [Undo]:
  ```
  (specify the third point, etc.)

- When you are done press [Enter] to end the WP mode.
- Whatever is inside the **WP** FULLY will be selected. If any part (even a tiny part) is outside the shape it will not be selected. See the example:

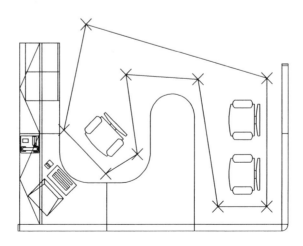

CP (Crossing Polygon)	■	**CP** is the same as **WP** except it has the features of **C**. Which means that whatever it fully contains, plus any object that it touches, will be selected.
F (Fence)	■	The main function of this mode is to touch objects.
	■	Discussed when we introduced Trim and Extend.
L (Last)	■	To select the last object drawn.
P (Previous)	■	To select the last selection set used.
All	■	To select all objects in the current file.
Deselect	■	If you select a group of objects, then you discover that one or two of the selected objects were selected by mistake, how do you deselect them?
	■	Simply hold the [Shift] key and click these objects, they will be deselected.

OTHER METHODS FOR SELECTING OBJECTS

- There are other methods for selecting objects, which will make your life easier.
- There is a nice technique called **Noun/Verb selection**, which will allow the user to select first, then issue the command.
- The cursor looks like the following:

- As you can see, there is a pick box inside the cursor.
- Which means, without issuing any command, you can:
  Click on any object to select it.
  - Or you can find an empty space, then click and go to the right to get the **Window** mode.
  - Or you can find an empty space, then click and go to the left to get the **Crossing** mode.

- Once you select the desired objects, you can right-click and you will get the following shortcut menu:

- From this shortcut menu you can count five modifying commands that you can access without typing a single letter on the keyboard, or issuing any commands either from menus, or from the toolbars. These commands are: **Erase**, **Move**, **Copy Selection**, **Scale**, and **Rotate**.
- Make sure that the **Noun/Verb** technique is checked on by selecting **Tools/Options** from the menu, selecting the **Selection** tab, and under **Selection Modes**, clicking the **Noun/Verb selection** checkbox to turn it on (if it is off).

- This technique will not work with the Offset, Fillet, Chamfer, Trim, Extend, and Lenghten commands.

MOVE COMMAND

- To move objects from one place to another.
- Use one of the following methods to issue the command:
  - From the dashboard and using **2D Draw** panel, click the **Move** button.

- From menus select **Modify/Move**.
- Type **move** (or **m**) in the Command window.
- This command is a three-step command.
- The first step is to:

```
Select objects:
```

- Once you are done press [Enter] or right-click.
- The next prompt will ask you to:

```
Specify base point or [Displacement] <Displacement>:
```
(Specify the base point)

- The ***Base point*** concept will be repeated in four other commands, so what is Base point?
    - The simplest definition of Base point is *Handle* point.
    - We don't have a golden rule, which will always define the right point for a Base point. Rather, you have to take it case by case. So it may sometimes be the center of a group of objects, and in another situation it will be the upper left corner.
    - This is true for commands like Move, Copy, and Stretch. But for a command like Rotate, Base point means the point which the whole shape will rotate around it. In Scale command, it will be the point that the whole shape will shrink or enlarge relative to it.
- The third prompt will be:

```
Specify second point or <use first point as displacement>:
```
(Specify the second point)

- The command will end automatically.
- See the following example:

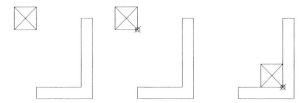

MOVING OBJECTS

Exercise 19

1. Start AutoCAD 2008.
2. Open the file **Exercise_19.dwg**.

3. Move the four objects (Bath Tub, Toilet, Sink, and Door) to their respective places to make the bathroom look like the following:

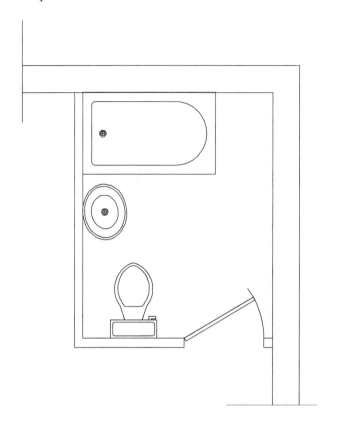

- For the Toilet use the midpoint of the inside wall.
- For the Sink use the quadrant of its left and the midpoint of the inside wall.

4. Save the file and close it.

COPY COMMAND

- To copy objects.
- Use one of the following methods to issue the command:
  - From the dashboard and using **2D Draw** panel, click the **Copy** button.

- From menus select **Modify/Copy**.
- Type **copy** (or **cp**) in the Command window.
- This command is a three-step command.
- The first step is to:

    Select objects:

- Once you are done press [Enter] or right-click.
- The next prompt will ask you to:

    Specify base point or [Displacement] <Displacement>: ***(Specify the base point)***

- The third prompt will be:

    Specify second point or [Exit/Undo] <Exit>: ***(Specify the second point)***
    Specify second point or [Exit/Undo] <Exit>: ***(Specify another second point)***

- Once you are done press [Enter] or right-click.
- If you made any mistakes simply use **u** to undo the last action.
- See the following example:

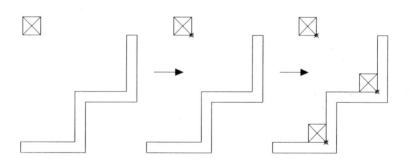

COPYING OBJECTS

Exercise 20

1. Start AutoCAD 2008.
2. Open the file **Exercise_20.dwg**.

3. Copy the chair three times to make the room look like the following:

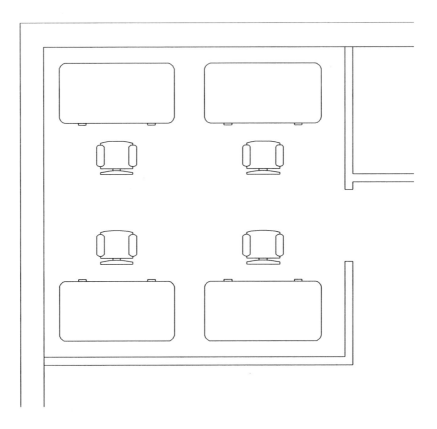

- Don't worry about the rotation of the chairs—we will fix this in the next exercise.

4. Save the file and close it.

ROTATE COMMAND

- To rotate objects around a point using a rotation angle.
- Use one of the following methods to issue the command:
  - From the dashboard and using **2D Draw** panel, click the **Rotate** button.
  - From menus select **Modify/Rotate**.
  - Type **rotate** (or **ro**) in the Command window.

- This command is a three-step command.
- The first step is to:

  `Select objects:`

- Once you are done press [Enter] or right-click.
- The next prompt will ask you to:

  `Specify base point:` ***(Specify the base point)***

- The third prompt will be:

  `Specify rotation angle or [Copy/Reference] <0>:`
  (specify the rotation angle, -=CW, +=CCW)

- You can use the **Copy** option if you want to rotate a copy of the objects selected while keeping the original intact.
- The command will end automatically.
- See the following example:

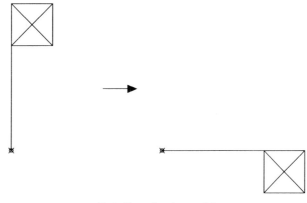

Rotation Angle = -90

ROTATING OBJECTS

 Exercise 21

1. Start AutoCAD 2008.
2. Open the file **Exercise_21.dwg**.

3. Rotate the two chairs (each in a different command) around the center of each chair to make the room look like the following:

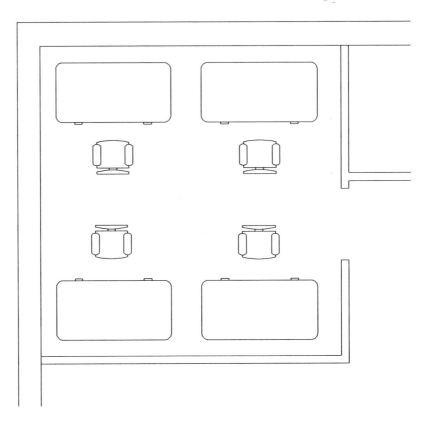

- The best way to do this exercise is by either using POLAR or typing the angles.
- You may need some movement to make the room look perfect.

4. Save the file and close it.

SCALE COMMAND

- To enlarge or shrink objects using a scale factor.
- Use one of the following methods to issue the command:
  - From the dashboard and using **2D Draw** panel, click on the small triangle at the right, keep holding, then click the **Scale** button.
  - From menus select **Modify/Scale**.
  - Type **scale** (or **sc**) in the Command window.

- This command is a three-step command.
- The first step is to:

  `Select objects:`

- Once you are done press [Enter] or right-click.
- The next prompt will ask you to:

  `Specify base point:` ***(Specify the base point)***

- The third prompt will be:

  `Specify scale factor or [Copy/Reference] <1.0000>:` ***(specify the scale factor, the number should be a non-zero positive number)***

- You can use the **Copy** option if you want to scale a copy of the objects selected while keeping the original intact.
- The command will end automatically.
- See the following example:

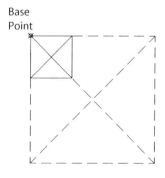

Base Point

Scalefactor=3

SCALING OBJECTS

Exercise 22

1. Start AutoCAD 2008.
2. Open the file **Exercise_22.dwg**.
3. Use scale factor = 0.8 to scale the bath tub using the upper left corner as the Base point.

4. Use scale factor = 1.2 to scale the sink using the quadrant of the left side as the Base point.
5. The room should look like the following:

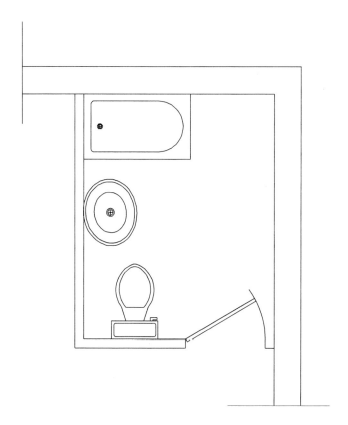

6. Save the file and close it.

ARRAY COMMAND

- To create a duplicate of objects using two methods:
  - Rectangular array (matrix shape)
  - Polar array (circular or semicircular shape)
- Use one of the following methods to issue the command:
  - From the dashboard and using **2D Draw** panel, click the **Array** button.

Rectangular
- From menus select **Modify/Array**.
- Type **array** (or **ar**) in the Command window.
- If you want to duplicate certain objects simulating the matrix shape, you want a rectangular array.
- The following dialog box will appear:

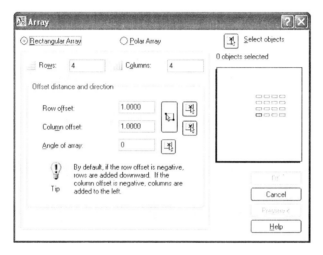

FIGURE 5-1

- First click the **Select Objects** button to select the desired objects. Once you are done press [Enter] or right-click.
- Specify the number of **Rows** and the number of **Columns** (original object is inclusive).
- Specify the **Row offset** (which is the distance between rows) and specify the **Column offset** (which is the distance between columns). While you are doing this keep two things in mind:
  - You have to be consistent. Measure the distance from the same reference (e.g., it is either from top-to-top, or from bottom-to-bottom, or from center-to-center, etc.).
  - You have to take care of the direction of copying. If you input a positive number, it will be either to the right or up. If you input negative number, it will be either to the left or down.
- Specify the **Angle of array**. By default it will be repeating the objects using the orthogonal angles.

- Click the **Preview** button to see the result of your inputs.

- AutoCAD will display the result with the following dialog box:

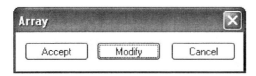

FIGURE 5-2

- If you like the results, click **Accept**. The command will end.
- If not, click **Modify**. The original dialog box will be back for further editing.
- See the following example:

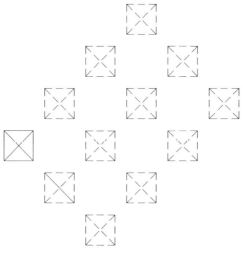

Dashed objects are duplicates, # of rows = 3,
# of columns = 4, row offset = –ve,
column offset = +ve, Angle of array = 45

RECTANGULAR ARRAY

Exercise 23

1. Start AutoCAD 2008.
2. Open the file **Exercise_23.dwg**.

3. Using a rectangular array, arrange the chairs to look like the following:

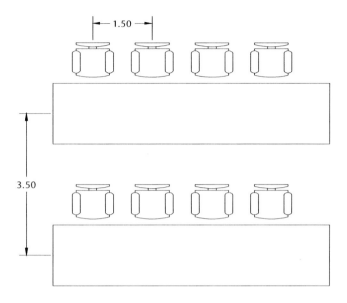

4. Save the file and close it.

Polar
- If you want to a duplicate certain objects simulating circular or semicircular shape, you want a Polar array.
- The following dialog box will appear:

FIGURE 5-3

- First click the **Select Objects** button to select the desired objects. Once you are done press [Enter] or right-click.

- Specify the **Center point** of the array, either by inputting the coordinates in X and Y, or by clicking the **Pick Center Point** button and specifying the point by mouse.
- You have three pieces of data to input. AutoCAD will only take two of them. These are:
  - Total number of items.
  - Angle to fill.
  - Angle between items.
- The following diagram will illustrate the relationship between the three parameters:

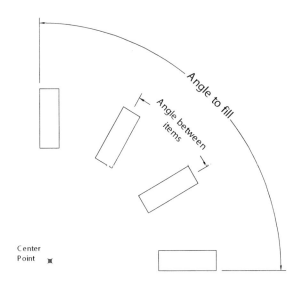

Total number of items = 4

- So you can specify two out of three parameters, which makes 3 different methods. They are:
  - Method 1: Specify the total number of items and Angle to fill. AutoCAD will figure out the angles between items.
  - Method 2: Specify the total number of items and Angle between items. AutoCAD will know the Angle to fill.
  - Method 3: Specify the Angle to fill and Angle between items. AutoCAD will calculate the total number of items.
- Under **Methods and values**, select the proper method and input the corresponding values.

- Specify if you want to **Rotate items as copied** or not. See the example below:

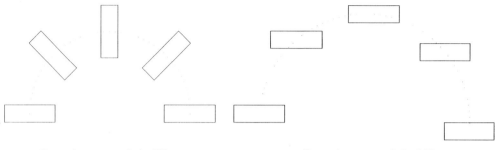

Rotate items as copied = ON   Rotate items as copied = OFF

POLAR ARRAY

Exercise 24

1. Start AutoCAD 2008.
2. Open the file **Exercise_24.dwg**.
3. Using the Polar array, arrange the square so you will get the following result:

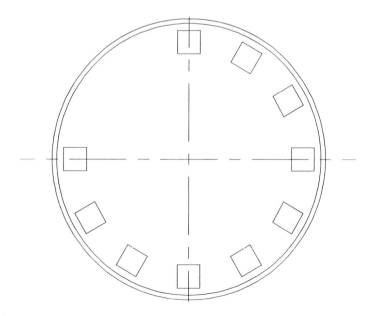

4. Save the file and close it.

MIRROR COMMAND

- To create a mirror image of selected objects.
- Use one of the following methods to issue the command:
  - From the dashboard and using **2D Draw** panel, click the **Mirror** button.
  - From menus select **Modify/Mirror**.
  - Type **mirror** (or **mi**) in the Command window.
- The first step is to:

  `Select objects:`

- Once you are done press [Enter] or right-click.
- Now you need to specify the mirror line by specifying two points:

  `Specify first point of mirror line:` *(specify the first point of the mirror line)*
  `Specify second point of mirror line:` *(specify the second point of the mirror line)*

- The following applies to the mirror line:
  - There is no need to draw a line to act as a mirror line—two points will do the job.
  - The length of the mirror is not important, but the location and angle of the mirror line will affect the final result.
- The last prompt will be:

  `Erase source objects? [Yes/No] <N>:` *(type N or Y)*

- Mirror command will produce an image in all cases, but what should AutoCAD do with the source objects? You can keep them or erase them.
- Mirror command ends automatically.

- If part of the objects to be mirrored is text, you have to decide whether you want to treat it as the other objects and mirror it or just copy it.
- To do that, prior to issuing the Mirror command, type in the Command window **mirrtext**. The following prompt will appear:

  `Enter new value for MIRRTEXT <0>:`

- If you input 0 (zero), then text will be copied.
- If you input 1, then text will be mirrored.
- See the following example:

MODIFYING COMMANDS   **117**

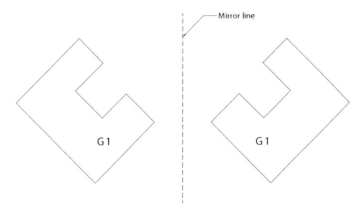

Mirrtext = 0 & Delete source objects = No

MIRRORING OBJECTS

 Exercise 25

1. Start AutoCAD 2008.
2. Open the file **Exercise_25.dwg**.
3. Using the Mirror command, produce the shape below:

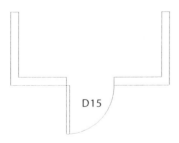

4. Save the file and close it.

STRETCH COMMAND

- To change the length of selected objects.
- Use one of the following methods to issue the command:
  - From the dashboard and using **2D Draw**, click the **Stretch** button.
  - From menus select **Modify/Stretch**.
  - Type **stretch** (or **s**) in the Command window.

- The first step is to:

  ```
  Select objects to stretch by crossing-window or
  crossing-polygon...
  ```
- The Stretch command is one of few commands that insists on a certain method of selecting.
- The Stretch command asks the user to select using either C or CP.
- As we discussed previously, C and CP will select any object contained inside and any object touched (crossed) by C or CP lines.
- The Stretch command will utilize both facilities by setting the following rules:
  - Any object contained FULLY inside C or CP will be moving.
  - Any object crossed by C or CP will be stretched.
  - Once you are done press [Enter] or right-click.
- The second prompt will be:

  ```
  Specify base point or [Displacement] <Displace-
  ment>:
  ``` *(specify Base point)*
- The third prompt will be:

  ```
  Specify second point or <use first point as dis-
  placement>:
  ``` *(specify the destination point)*
- The Stretch command ends automatically.
- See the following example:

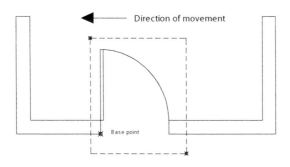

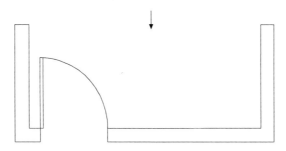

MODIFYING COMMANDS **119**

STRETCHING OBJECTS

Exercise 26

1. Start AutoCAD 2008.
2. Open the file **Exercise_26.dwg**.
3. Using the Stretch command, move the door 2 units to the left, so it will look like the following:

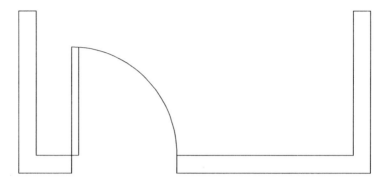

4. Save and close the file.

BREAK COMMAND

- To break an object into two pieces.
- Use one of the following methods to issue the command:
    - From the dashboard and using **2D Draw** panel, click on the small triangle at the right, keep holding, then click the **Break** button.
    - From menus select **Modify/Break**.
    - Type **break** (or **br**) in the Command window.
- The first step is to:

  ```
  Select object:
  ```

- You can break one object at a time. When you select this object, AutoCAD will issue you the following prompt:

  ```
  Specify second break point or [First point]:
  ```

- To understand this prompt, take care of the following points:
    - In order to break an object you have to specify two points on it.
    - The selecting you made can be considered a selection and also a first point, or can be considered a selection only. If you consider the

selection is a selection and a first point, respond to this prompt by specifying the second point.
- On the other hand, if you want the selection to be only a selection, type the letter F and AutoCAD will respond by the following prompt:

```
Specify first break point: (specify first break-
ing point)
Specify second break point: (specify second
breaking point)
```

- When you want to break a circle, take care to specify the two points CCW.
- See the following example:

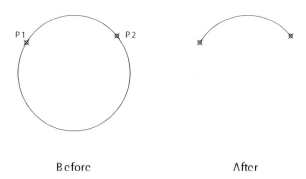

Before   After

- In dashboard and using **2D Draw** panel, there is another tool called **Break at Point**, which is similar to the **Break** command except the following:
  - You will be asked to select only one point.
  - AutoCAD will assume that the first point and the second point are in the same place.
  - The object will be broken into two objects, but connected.

BREAKING OBJECTS

Exercise 27

1. Start AutoCAD 2008.
2. Open the file **Exercise_27.dwg**.

3. Using the Break command, break the two circles to look something like:

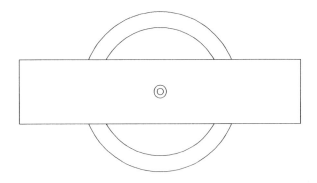

4. Save the file and close it.

GRIPS: INTRODUCTION

- Grips is a method of modifying your objects in an easy and fast fashion.
- Grips is a simple click on object(s) without issuing any command.
- It will do two things for you:
  - It will select the objects, hence they will be ready for any Modifying commands to be issued, and they will be the Selection Set for this command (discussed at the beginning of this Chapter).
  - A blue (default color) square will appear on certain places dependent on the type of object. Here are some examples:

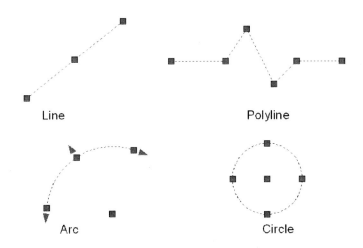

- These squares are the grips.
- There is a magnet relationship between these squares and the pick box of the croshairs. So if you hover over them, the blue will turn to green to indicate that this is the current grip right now.
- If you click on one of these blue squares, you will:
  - Make it hot, turning it to red.
  - Make this grip a base point.
  - Start a group of five Modify commands (use the right-click).

GRIPS: THE FIVE COMMANDS

- Once you make one of the blue squares hot (by clicking on it), this grip will become the Base point for five commands. They are:
  - Move
  - Mirror
  - Rotate
  - Scale
  - Stretch
- To see these commands, right-click, and the following shortcut menu will appear:

- As you can see, other options available in the shortcut menu are:
  - **Base Point**, which is to define a new Base point other than the one you start with.

- **Copy**, means a mode rather than to be a command. Copy mode works with all other five commands mentioned above, and hence, this will give you the ability to Rotate with Copy, Scale with Copy, etc.

GRIPS: STEPS & NOTES

- The steps to use Grips are as follows:
  - Select the object(s) desired (direct clicking, window mode, or crossing mode).
  - Select one of the grips to be your Base point, and click it. It will become hot (by default red).
  - Right-click and select from the shortcut menu the desired command. Meanwhile, you can specify another Base point, and/or you can select the Copy mode.
  - Do the steps of the desired command.
  - Once you are done press [Esc] once.
- You can use OSNAP with Grips with no limitations. Also, you can use Polar and OTRACK, so the modification will be done with high accuracy.
- Mirror is the only command that doesn't ask explicitly for a Base point. So how come it is listed with the other four commands? AutoCAD considers the first point of the mirror line to be the Base point.
- Using Grips, to keep both the original and the mirrored image you have to select the Copy mode after you select the Mirror command.
- You can *deselect* certain objects from the Grips by holding [Shift] and clicking on this object, avoiding the squares.

GRIPS AND DYN

- DYN has a facility for giving information about the objects with their grips appearing on the screen.
- If you hover over an end grip of a line, DYN tells the length and the angle of that line.

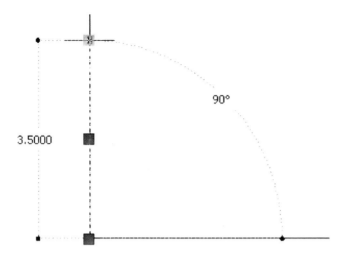

- If you hover over an end grip shared between two lines, DYN tells the length and angle of both lines.

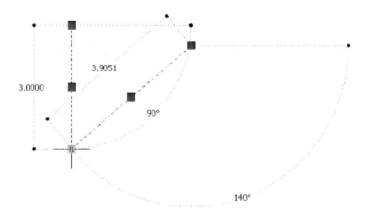

- If you hover over the middle grip of an arc, DYN tells the radius and the length of the arc.

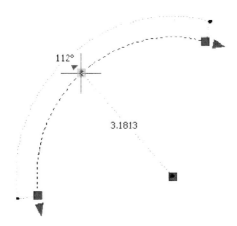

- If you hover over the quadrant grip of a circle, DYN tells the radius of the circle.

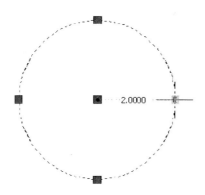

USING GRIPS

Exercise 28

1. Start AutoCAD 2008.
2. Open the file **Exercise_28.dwg**.
3. Without issuing any commands, select the upper circle. Make the center hot, then right-click and select **Scale**, then right-click again and select **Copy**. For the **Scale factor** prompt type **0.5**. Press [Esc] twice.
4. In the right part of the base, without issuing any commands, select the rectangle. Make one of the blue grips hot by clicking it, right-click and select **Rotate**, then right-click again and select **Base Point** so you can

specify a new base point, which is the center of the rectangle (using OSNAP and OTRACK), and use **Rotation angle** = 90. Press [Esc] twice.
5. Select the rotated rectangle at the right part of the base. Select any grip to make it hot, then right-click and select **Mirror**, then right-click again and select **Copy**, and right-click for the thrid time and select **Base Point**, and specify one of the two endpoints of the vertical lines seperating the two parts of the base, then specify the other endpoint. Press [Esc] twice. The shape will end up like:

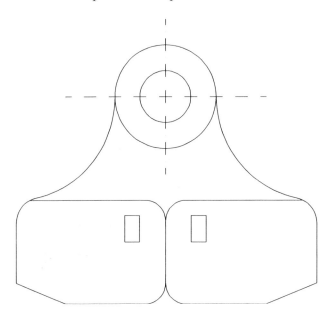

6. Save the file and Close.

CHAPTER REVIEW

1. Move, Copy, Rotate, Scale, Stretch, what is common between these commands?
   a. They are all Modifying commands.
   b. They all use the Base point concept.
   c. They all change the length of an object.
   d. A & B.
2. In Stretch command, you have to use _____ or _____ while selecting the objects.

3. Mirrtext is to control whether to copy or to mirror the text in Mirror command:
   a. True
   b. False
4. If you break a circle, take care to specify the two points:
   a. CCW.
   b. CW.
   c. Doesn't matter.
   d. You can't break a circle.
5. You can scale using Scale factor = -1
   a. True
   b. False
6. In the Array command using the Rectangular option, the Row offset must be _____ if you want to repeat the objects downward.

CHAPTER REVIEW ANSWERS

1. d
2. C or CP
3. a
4. a
5. b
6. negative

Chapter 6

DEALING WITH BLOCKS

In This Chapter
- What are blocks?
- Creating and inserting blocks
- Exploding blocks
- Design Center and Tool Palettes and their effect on blocks
- Editing blocks

WHAT ARE BLOCKS?

- A block in AutoCAD is any shape that is repeated in one or more drawings more than once.
- So, instead of drawing it each and every time you need it, follow the following steps:
  - Draw it once.
  - Store it as a block.
  - Insert it as many times as you wish.
- Blocks in AutoCAD changed a lot in the past five years, which made some old procedures obsolete.
- In our discussion we will mention the old methods, but we will concentrate more on the new methods of using blocks.

CREATING BLOCKS

- The first step in creating blocks in AutoCAD is to draw the desired shape.
- While drawing the shape, consider the following three guidelines:
  - Draw the shape in Layer 0 (zero).
  - Draw the shape in certain units (to be usable in Design Center).
  - Draw the shape in the right dimensions.

Why Layer 0?	• Layer 0 is different from any other layer in AutoCAD. It will allow the block to be transparent both in color and in linetype.
	• If you draw the shape while layer 0 is current, then insert it in another layer with red color and dash-dot linetype, the block will be red and dash-dot.
Why Certain Units?	• If you want to use your block with Design Center you need to specify the units you used to create it.
	• That will affect the Automatic scaling feature of Design Center.
What Are Right Dimensions?	• Right dimensions are either:
	• The real dimension of the shape.
	• Or, values, like 1, 10, 100, or 1000 for the distances, so it will be easy for you to scale the block once you insert it.
	• Let's assume we draw the following shape:

- The next step would be to think about a point, which will act as the Base point (the handle for this block, which you will carry once inserting it).
- Also, think about a good name for this block.
- If all of these things are ready in your mind, issue the command using one of the following ways:

    • From the dashboard and using **2D Draw** panel, click on the small triangle at the right, keep holding, then click the **Make Block** button.
    • From menus select **Draw/Block/Make**.
    • Type **block** (or **b**) in the Command window.

- The following dialog box will appear:

FIGURE 6-1

- Type the name of the block (it should not exceed 255 characters—similar to the layer naming conditions).

- Under **Base point**, click the **Pick point** button to input the base point of the block. Once you are done press [Enter] or right-click. Or, you can select the checkbox **Specify On-screen** to specify the base point after the dialog box closes.

- Under **Objects**, click the **Select objects** button to select the objects, which will form the block. Once you are done press [Enter] or right-click. Or, you can select the checkbox **Specify On-screen** to specify the base point after the dialog box closes.
- Now select one of three choices of what to do with the objects you draw to create this block from:
  - **Retain** them as objects.
  - **Convert** them **to block**.
  - **Delete** them.
- Leave **Annotative** off (this is an advanced feature).
- Select whether the block should always be **Scaled uniformly** (Xscale = Yscale) or not?
- Select whether the block can be exploded in the future or not?
- Under **Block unit**, select the unit you draw the shape in. This will help AutoCAD in the Automatic scaling feature in Design Center.
- Write any description for your block.
- Select whether to allow this block to be opened in the block editor (block editor is an advanced feature to create dynamic blocks). Keep it off for the time being.
- When you are done, click **OK**.
- At this moment, let's imagine that our drawing has a cabinet, the door of the cabinet will be opened, and the defined block will be put inside it. This block will be intact. Even when you insert it, you will insert an incidence of it.
- The fragmented objects become one object.
- You can define as many blocks as you wish.

CREATING A BLOCK (METRIC)

Workshop 3-A

1. Start AutoCAD 2008, and open the file: **Small_Villa_Ground_Floor_Plan_Metric_Workshop_3.dwg**. (If you solved the previous workshop correctly, open your file.)

2. Make layer 0 current.
3. Choose an empty space and draw the following shape (without dimensions):

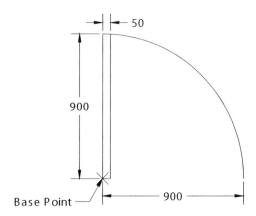

4. Using the Make Block command, create a block using the following information:
   a. Block name = Door.
   b. Specify the designated base point.
   c. Delete the shape after the creation of the block.
   d. Block unit = Millimeters.
   e. Scale uniformly = off, Allow exploding = on.
   f. Description = Door to be used inside the building. Refer to the door table.
   g. Open in block editor = off.
5. Save the file and close it.

CREATING A BLOCK (IMPERIAL)

 Workshop 3-B

1. Start AutoCAD 2008, and open the file: **Small_Villa_Ground_Floor_Plan_Imperial_Workshop_3. dwg**. (If you solved the previous workshop correctly, open your file.)
2. Make layer 0 current.

3. Choose an empty space and draw the following shape (without dimensions):

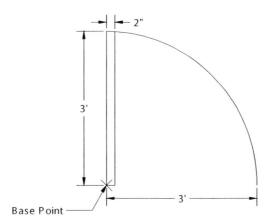

4. Using the Make Block command, create a block using the following information:
   a. Block name = Door.
   b. Specify the designated base point.
   c. Delete the shape after the creation of the block.
   d. Block unit = Inches.
   e. Scale uniformly = off, Allow exploding = on.
   f. Description = Door to be used inside the building. Refer to the door table.
   g. Open in block editor = off.
5. Save the file and close it.

INSERTING BLOCKS

- Once you create a block, you can use it in your drawing as many times as you wish.
- Inserting the block in your drawing requires you to consider the following guidelines:
  - Set the current layer to be the layer that the block belongs to.
  - You have to make the drawing ready to accommodate the block (e.g., finish the door openings before inserting the door block).
- Now you can issue the command using one of the following ways:
  - From the dashboard and using **2D Draw** panel, click on the small triangle at the right, keep holding, then click the **Insert Block** button.

- From menus select **Insert/Block**.
- Type **insert** (or **i**) in the Command window.
- The following dialog box will appear:

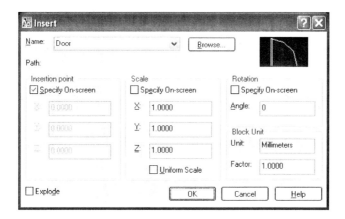

FIGURE 6-2

- Select the name of the desired block from the list.
- Specify the **Insertion point** using one of two methods:
  - Clicking the **Specify On-screen** checkbox on, which means you will specify the insertion point using the mouse (most likely this is handier than the next method).
  - Typing the coordinates of the insertion point.
- Specify the **Scale** of the block by using one of the following methods:
  - Clicking the **Specify On-screen** checkbox on, which means you will specify the scale using the mouse.
  - Typing the scale factor in all three directions of the insertion point, which means you can set the X scale factor not equal to the Y scale factor.
  - Another way to do scaling is by clicking the **Uniform Scale** checkbox on, which will allow you to input only one scale. The others will follow.
- Specify the **Rotation** of the block by using one of the following methods:
  - Clicking the **Specify On-screen** checkbox on, which means you will specify the rotation angle using the mouse.
  - Typing the rotation angle.
- The **Block unit** part will be read-only, and hence will show you the unit that you specified when you created this block. Also, it will show the **Factor** that is based on the Block unit and the drawing unit (which is defined in the **Format/Units** dialog box).

- Click **OK**, to end the command.
- Using the Scale of the block, you can use negative values to insert mirror images of your block.
- See the following example:

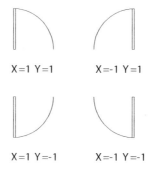

INSERTING BLOCKS (METRIC)

 Workshop 4-A

1. Start AutoCAD 2008, and open the file: **Small_Villa_Ground_Floor_Plan_Metric_Workshop_4. dwg**. (If you solved the previous workshop correctly, open your file.)
2. Make the **Doors** layer current.
3. Using the **Insert** command, insert block **Door** in the proper places as shown below:

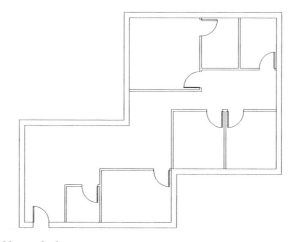

4. Save the file and close it.

INSERTING BLOCKS (IMPERIAL)

 Workshop 4-B

1. Start AutoCAD 2008, and open the file: **Small_Villa_Ground_Floor_Plan_Imperial_Workshop_4. dwg**. (If you solved the previous workshop correctly, open your file.)
2. Make the **Doors** layer current.
3. Using the **Insert** command, insert block **Door** in the proper places as shown below:

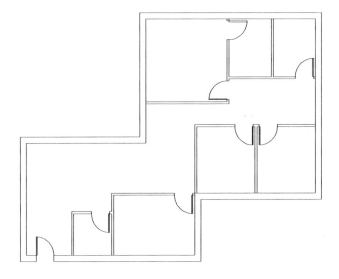

4. Save the file and close it.

EXPLODING BLOCKS

- By default when you insert incidences of blocks, keep them as blocks, and don't try to change their nature. But, in some (rare) cases, you may want to explode the block (which is one object) to the objects forming it.
- You have to use the Explode command.
- To issue the command try one of the following methods:

  - From the dashboard and using **2D Draw** panel, click on the small triangle at the right, keep holding, then click the **Explode** button.

- From menus select **Modify/Explode**.
- Type **explode** (or **x**) at the Command window.
- In all cases, AutoCAD will prompt you to:

```
Select objects:
```

- Once you are done, press [Enter] or right-click. Once you explode a block it will go back to its original layer (the layer the block was created in).

- Don't use this command unless you really need it.
- In older versions of AutoCAD a block can't be defined as a cutting edge or a boundary edge in both the Trim and Extend command. In AutoCAD 2005, it was allowed to select blocks as cutting edges and boundary edges.
- You can use this command to explode **Pline** to lines and arcs.

SHARING DATA BETWEEN AUTOCAD FILES USING DESIGN CENTER (BLOCKS)

- Before AutoCAD 2000, if you created a block in one drawing you couldn't use it in other drawings unless you converted it to a file.
- In AutoCAD 2000, Autodesk introduced a very nice tool called **Design Center**, which will allow the user to share blocks, layers, and other things between different files.
- We will concentrate on blocks here.
- The file you want to take the blocks from could be anywhere:
  - It can be in your computer.
  - It can be in your colleague's computer, which is hooked up on the local area network of your company.
  - It can be in one of the websites on the Internet.
- To start the Design Center command, use one of the following methods:

  - From the **Standard Annotation** toolbar, click the **Design Center** button.
  - From menus select **Tools/Design Center**.
  - Type **adcenter** in the Command window.
  - Press **[Ctrl] + 2** (number 2 on the upper part of the keyboard and not the righthand side).

- The following will appear on the screen:

FIGURE 6-3

- As you can see, the Design Center palette is split into two parts.
- On the left you will see the hierarchy of your computer, including all your hard disks and network places (just like **My Computer** in Windows).
- Select the desired hard disk and double-click it. A list of folders will appear. Select the desired folder and double-click it. You should see a list of drawings. Select the desired drawing and double-click it. You will see the following:

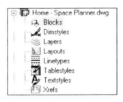

- As you can see from above, you can take from this drawing the following:
  - Blocks
  - Dimstyles
  - Layers
  - Layouts
  - Linetypes
  - Tablestyles
  - Textstyles
  - Xrefs

- We will look at Blocks here. Once you click (you don't need to double-click) the word Blocks, look at the right part of Design Center. You will see the blocks available in this drawing.
- There are several ways to take blocks from this drawing to your drawing. They are:
  - Drag-and-Drop (using the left button)
  - Drag-and-Drop (using the right button)
  - Double-click
  - Right-click

Drag-and-Drop Using the Left Button
- Make sure that you are in the right layer.
- Make sure that you switched on the right OSNAP settings.
- Click and hold on the desired block.
- Drag it into your drawing—you will be holding it from the Base point.
- Once the right OSNAP is caught release the mouse button.

Drag-and-Drop Using the Right Button
- Make sure that you are in the right layer.
- Make sure that you switched on the right OSNAP settings.
- Right-click and hold on the desired block.
- Drag it into your drawing.
- Release the mouse button. The following shortcut menu will appear:

- Select **Insert Block**. The following dialogue box will appear:

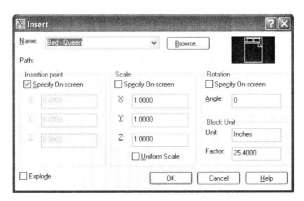

FIGURE 6-4

- It is the same as the **Insert** dialog box.
- Do the same steps as discussed before.

Double-click
- If you double-click any block, the **Insert** dialog box will appear.

Right-click
- Select the desired block and right-click. The following shortcut menu will appear:

- If you select **Insert Block**, the Insert dialog box will appear as discussed before.
- As for the two options **Insert and Redefine** and **Redefine only**, we will be discussing redefining in the next pages.
- Select the **Block Editor** option if you want to open this block in the Block Editor in order to add Dynamic features to it.
- If you select **Copy**, that means you will copy to the clipboard of Windows, and hence you can use it in AutoCAD or other software. To use it select **Edit/Paste** or **Ctrl+V**.
- In the coming pages we will discuss **Tool Palette**, hence we will discuss the last option in this shortcut menu.

BLOCK AUTOMATIC SCALING

- If you are using the Design Center to bring some blocks from other drawings and find that the block is either too big or too small, you will know that there is something wrong with the Automatic Scaling.
- To control the Automatic Scaling you have to do the following two steps:
  - While you are creating the block, make sure you are setting the right **Block unit**.
  - Before you bring the block from Design Center, set the **Units to scale inserted contents** in **Format/Units**.

Block unit
- When you are creating a block, the following dialog box will appear:

FIGURE 6-5

Units to Scale Inserted Contents
- Under the part labeled **Block unit**, select the desired unit.
- Before using any block from the Design Center, select **Format/Units**. The following dialog box will appear:

FIGURE 6-6

- Under the part labeled **Units to scale inserted contents**, set the desired scale used in your drawing.
- Using the two scales AutoCAD will calculate the proper scale the Block should appear with.

USING DESIGN CENTER (METRIC)

 Workshop 5-A

1. Start AutoCAD 2008, and open the file: **Small_Villa_Ground_Floor_Plan_Metric_Workshop_5. dwg**. (If you solved the previous workshop correctly, open your file.)
2. Make the **Furniture** layer current.
3. Select **Format/Units** and make sure that **Units to scale inserted content** is **Millimeters**.
4. Open the Design Center. From the left part of the Design Center Pallette double-click the Drive containing the AutoCAD 2008 folder.
5. Select **AutoCAD 2008/Sample/Design Center**.
6. Using **Home Space Planner.dwg**, **House Designer.dwg** and **Kitchens.dwg** while OSNAP is off, drag-and-drop the following blocks as shown:

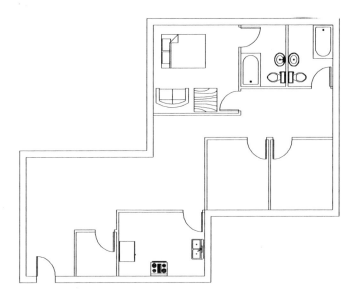

7. Save the file and close it.

USING DESIGN CENTER (IMPERIAL)

 Workshop 5-B

1. Start AutoCAD 2008, and open the file: **Small_Villa_Ground_Floor_Plan_Imperial_Workshop_5.dwg**. (If you solved the previous workshop correctly, open your file.)
2. Make the **Furniture** layer current.
3. Select **Format/Units** and make sure that **Units to scale inserted content** is **Inches**.
4. Open the Design Center. From the left part of the Design Center Pallette double-click the Drive containing the AutoCAD 2008 folder.
5. Select **AutoCAD 2008/Sample/Design Center**.
6. Using **Home Space Planner.dwg**, **House Designer.dwg** and **Kitchens.dwg** while OSNAP is off, drag-and-drop the following blocks as shown:

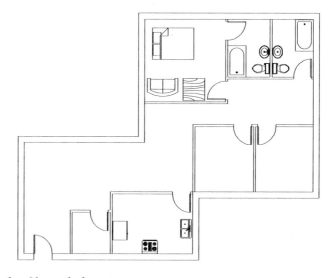

7. Save the file and close it.

TOOL PALETTES: INTRODUCTION

- This is a relatively new feature in AutoCAD, first introduced in AutoCAD 2004, and enhanced in AutoCAD 2005. When you are using Design Center, you will discover that you will need to locate the folder and the drawing that has the desired blocks you need.

- So one of the demands will be to keep these blocks available all the time. To do that use Tool Palettes.
- Tool Palette will keep Blocks, Hatch, and other stuff available for you regardless of which drawing you in are right now.
- In AutoCAD 2004, only Blocks and Hatch were available, but in AutoCAD 2005 you can keep virtually anything in it.
- Tool Palette works with the same Drag-and-Drop method we learned in Design Center. But in Tool Palette there are two methods—from Tool Palette and to Tool Palette. We will discuss them in the coming pages.
- Tool Palette is unique per computer and not per drawing, hence if you create (customize) a Tool Palette it will be available for all your drawings.
- To start the Tool Palette command, use one of the following methods:
    - From the **Standard Annotation** toolbar, click the **Tool Palettes Window** button.
    - From menus select **Tools/Palettes/Tool Palettes**.
    - Type **toolpalettes** in the Command window.
    - Press **[Ctrl] + 3** (number 3 at the upper part of the keyboard, not the right side).
- The following will appear on the screen:

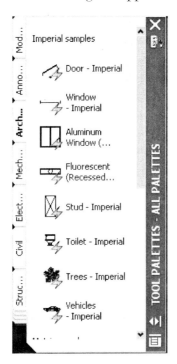

FIGURE 6-7

- You will see that several Tool Palettes are premade by AutoCAD for your immidiate use.
- You can create your own Tool Palette using different methods, depending on the source.
- You can copy, cut, and paste Tools inside each Tool Palette.
- You can customize the Tools inside each Tool Palette.

CREATING TOOL PALETTES FROM SCRATCH

- Right-click over the name of any existing Tool Palette and the following shortcut menu will appear:

- Select the **New Palette** option and a new Tool Palette will be added. Type in the name of the new Tool Palette, just like:

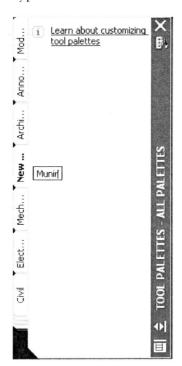

FIGURE 6-8

- A new empty Tool Palette is added.
- How do you fill this Tool Palette? Use the Drag-and-Drop method, whether from the graphical screen or from Design Center.

NOTE
- By default the local blocks of the current drawing are *NOT* automatically available in your Tool Palettes.

Example
- Assume we have the following drawing in front of us. This drawing contains lines, hatch, and dimensions:

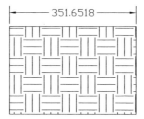

- Without issuing any commands, click on the line then hold (avoid the blue rectangles) and drag it into the empty Tool Palette. Do the same thing for the hatch and the linear dimension. Your Tool Palette will look like:

FIGURE 6-9

CREATING TOOL PALETTES USING DESIGN CENTER

- You can copy all blocks in one drawing using Desing Center and create a Tool Palette from them, keeping the name of the drawing.
- To do that, follow these steps:
  - Start Design Center.
  - Go to the desired folder, then to the desired file.
  - Right-click on the **Blocks** icon and the following shortcut menu will appear:

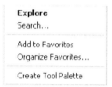

- Select the **Create Tool Palette** option, a new Tool Paltte will be added keeping the name of the file you chose and containg all the blocks.

- You can also drag-and-drop any block from the Design Center to any desired Tool Palette.

CUSTOMIZING TOOL PALETTES

- Blocks and Hatch patterns in a single Tool Palette can be copied and pasted in the same Tool Palette.
- The purpose of this is to assign different properties to each Tool.
- For example, there is a block called chair. You want to make 3 additional copies from it, each with a different rotation angle. The same thing applies to Hatch patterns as each copy can have a different scale factor.
- Also, you can specify that a certain block (or hatch) go to a certain layer regardless of what the current layer is.

How to copy a Tool
- Follow these steps:
  - Righ-click on the desired Tool and the following shortcut menu will appear:

How to Paste a Tool

- Select the **Copy** option.
- Follow these steps:
  - Select an empty area and right-click and the following shortcut menu will appear:

  - Select the **Paste** option. The copied Tool will reside at the bottom of the Tool Palette and will hold the same name.

How to Customize a Tool

- Follow these steps:
  - Right-click on the copied Tool and the following shortcut menu will appear:

  - Select the **Properties** option and the following dialog box will appear:

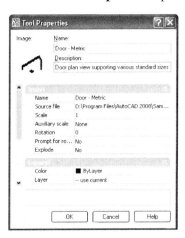

FIGURE 6-10

- You can change the **Name** and **Description** of the Tool.
- There are two types of properties:
  - Specific Tool properties; Block's properties are different from Hatch's properties.
  - General properties like, Color, Layer, Linetype, Plot style, and Lineweight.
- By default, the General properties are all **use current**, which means use the current settings.
- Both Specific Tool properties and General properies are all changeable.
- Go to the **General** and you will see something similar to the following:

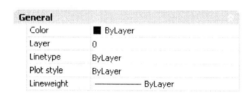

- Change the desired property you wish to change and then click **OK**.
- The following features apply to both Design Center and Tool Palettes:
  - Auto-hide feature. Both Design Center and Tool Palettes normally occupy space from the Graphical screen, hence making drafting more difficult. You can switch the Auto-hide feature on, so whenever the mouse pointer is away from either one of them only the title bar will appear. See the following illustration:

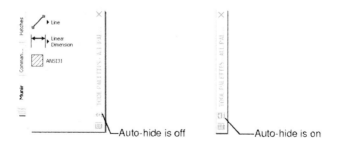

  - Both Design Center and Tool Palettes can be resized horizontally and vertically.
  - The user can specify to allow the docking of both Design Center and Tool Palettes on the right or left parts of the screen, or to keep them floating in the screen all the time.

■ In the **Tool Properties** dialogue box, if the image of the block is not clear you can change it. Do the following steps:
- Go to the image.
- Right-click. A small menu will appear with one choice—**Specify image**. Select this choice and the following dialog box will appear:

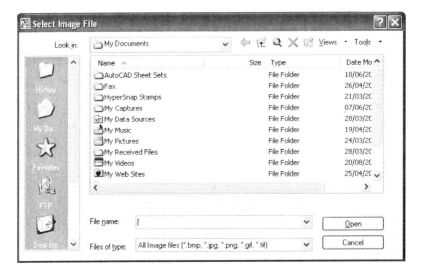

FIGURE 6-11

- Select the desired folder and file and click **Open**. Now the new image will apper.

USING AND CUSTOMIZING TOOL PALETTES (METRIC & IMPERIAL)

 Workshop 6-A & 6-B

1. Start AutoCAD 2008, and start a new file.
2. Make layer 0 current.
3. Open the Design Center. From the left part of the Design Center Pallette double-click the Drive containing the AutoCAD 2008 folder.
4. Select **AutoCAD 2008/Sample/Design Center**.
5. Select the file **Home Space Planner.dwg**.

6. Right-click the Blocks icon and select **Create Tool Palette**. A new Tool Palette with the name **Home Space Planner** will be added.
7. Select the file **House Designer.dwg**.
8. Show the blocks of this file.
9. Locate the **Bath Tub – 26 x 60 in** block, drag and drop it into your newly made Tool Palette.
10. Do the same thing with the **Sink Ovel Top** block.
11. Right-click on the name of the Tool Palette, select **Rename**, and change the name to **My Tools**.
12. In the **My Tools** Tool Palette, right-click on the Tool named **Chair – Rocking** and select **Properties**. In the dialog box change the **Layer** from **use current** to **Furniture**.
13. Right-click on the same Tool again and select **Copy**. Select an empty space and **Paste** it *three* times.
14. Using the same method in (12), change the Rotation angle of these blocks to 90, −90, and 180.
15. Now you have a tool pallete that you can use in all of your drawings on this computer.

EDITING BLOCKS

- Assume that after you created a block and inserted it several times in your drawing you discovered that there is something wrong with it.
- To solve your problem you need to redefine the orginal block.
- There are several methods to do that, some new, and some old.
- We will concentrate on the new method.
- In AutoCAD 2000, a new command was introduced called **Refedit**, which enables the user to **edit** the block definition **in place** without the need to insert, explode, and redefine (which is the old method).
  - From the **Refedit** toolbar, click the **Edit Reference In-Place** button.
  - From menus select **Tools/Xref and Block Editing/Edit Reference In-Place**.
  - Type **refedit** in the Command window.
- In all cases, AutoCAD will prompt you to:

```
Select reference:
```

- Select any incidence of the block. The following dialog box will appear:

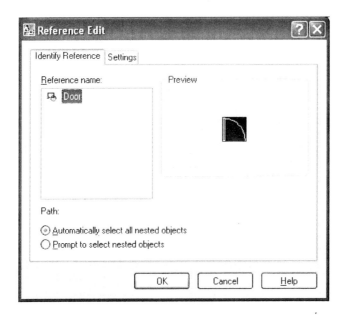

FIGURE 6-12

- Click **OK**. Once you do that, all of the drawing will be dimmed (you can't edit the dimmed objects), except your block.
- If, for any reason, you wanted other objects (normal objects and not blocks) available to you so you can edit the desired block, use the following:

- From the **Refedit** toolbar, click **Add to Working set**. AutoCAD will prompt:

```
Select objects:
```

- Select the desired objects to be added.
- The same thing if you want to remove some objects from the working set, use the following command:

- From the **Refedit** toolbar, click **Remove from working set**. AutoCAD will prompt:

```
Select objects:
```

- Select the desired objects to be removed.

- After you do your editing, you can either discard the changes as if nothing happened, or save back changes to the original block. Use either one of these commands:

- From the **Refedit** toolbar, click **Close Reference**. The following dialog box will appear:

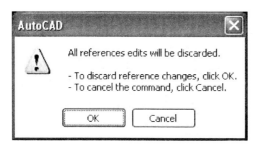

FIGURE 6-13

- If you really want to discard all the changes you made to the block, simply click **OK**, otherwise click **Cancel**.
- From the **Refedit** toolbar, click **Save Reference Edits**. The following dialog box will appear:

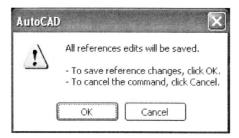

FIGURE 6-14

- Click **OK** if your are sure of the changes you made, otherwise click **Cancel**.
- If you issue the command **Save Reference Edits** and then you click **OK**, all indecedences of the block will change accordingly.
- Another way to issue the **Refedit** command is to click any incedence of the desired block, then right-click, and from the shortcut menu select **Edit Block In-Place**.

EDITING BLOCKS (METRIC)

 Workshop 7-A

1. Start AutoCAD 2008, and open the file: **Small_Villa_Ground_Floor_Plan_Metric_Workshop_7.dwg**. (If you solved the previous workshop correctly, open your file.)
2. Show the **Refedit** toolbar and start the **Refedit** command.
3. Select one of the doors you inserted and once the dialog box appears click **OK**. All of the drawing will be dimmed except the selected block.
4. Select the arc representing the swing of the door. Right-click and select **Properties**.
5. Change its layer to **Door_Swing**.
6. Change the Linetype scale = **200**.
7. Close the Properties palette.
8. Click on **Save back changes to reference**. At the dialog box click OK.
9. You will now see that all the door swings changed to dashed linetype.
10. Save the file and close it.

EDITING BLOCKS (IMPERIAL)

 Workshop 7-B

1. Start AutoCAD 2008, and open the file: **Small_Villa_Ground_Floor_Plan_Imperial_Workshop_7.dwg**. (If you solved the previous workshop correctly, open your file.)
2. Show the **Refedit** toolbar and start the **Refedit** command.
3. Select one of the doors you inserted and once the dialog box appears click **OK**. All of the drawing will be dimmed except the selected block.
4. Select the arc representing the swing of the door. Right-click and select **Properties**.
5. Change its layer to **Door_Swing**.
6. Change the Linetype scale = **10**.
7. Close the Properties palette.
8. Click on **Save back changes to reference**. At the dialog box click OK.
9. You will now see that all the door swings changed to dashed linetype.
10. Save the file and close it.

CHAPTER REVIEW

1. You should draw your orgianl shape, which will be a block, in layer 0:
   a. True
   b. False
2. Automatic scaling in Design Center works only with the _____ method.
3. The old method of editing a block is Refedit:
   a. True
   b. False
4. Which is true about Tool Palettes:
   a. Can be created from blocks coming from Design Center.
   b. You can Drag-and-Drop from and to a Tool Palette.
   c. You can put objects, hatches, dimensions, and blocks in Tool Palette.
   d. All of the above.
5. One of these commands is not for blocks:
   a. Explode command.
   b. Insert command.
   c. Makeblock command.
   d. Refedit command.
6. In order to make Design Center and Tool Palettes occupy less space on the graphical screen use _____.

CHAPTER REVIEW ANSWERS

1. a
2. Drag-and-Drop
3. b
4. d
5. c
6. Auto-hide

Chapter 7

HATCHING

In This Chapter

- Hatching in AutoCAD
- The old method of hatching using the Bhatch command
- The new method of hatching using Tool Palettes
- Gradient command
- Editing hatching in AutoCAD

HATCHING IN AUTOCAD

- Starting from AutoCAD 2005, it became acceptable to hatch an area with a small opening. Before this it was a prerequisite to be fully closed.
- AutoCAD comes with a good number of generic, predefined hatch patterns saved in a file called *acad.pat*. You can also buy other hatch patterns from third-party vendors, who can be found on the Internet.
- Hatch, like any other object, should be placed in a separate layer.
- There are two methods to hatch in AutoCAD, an old one (Bhatch command) and a new one (Tool Palette). We will start with the old one.

BHATCH COMMAND: SELECTING THE HATCH PATTERN

- This is the old method of hatching in AutoCAD.
- Issue the command using one of the following ways:

  - From the dashboard and using the **2D Draw** toolbar, locate the small triangle on the right, click and hold it, then select the **Hatch** button.
  - From menus select **Draw/Hatch**.
  - Type **bhatch** (or **bh**) in the Command window.

- The following dialog box will appear:

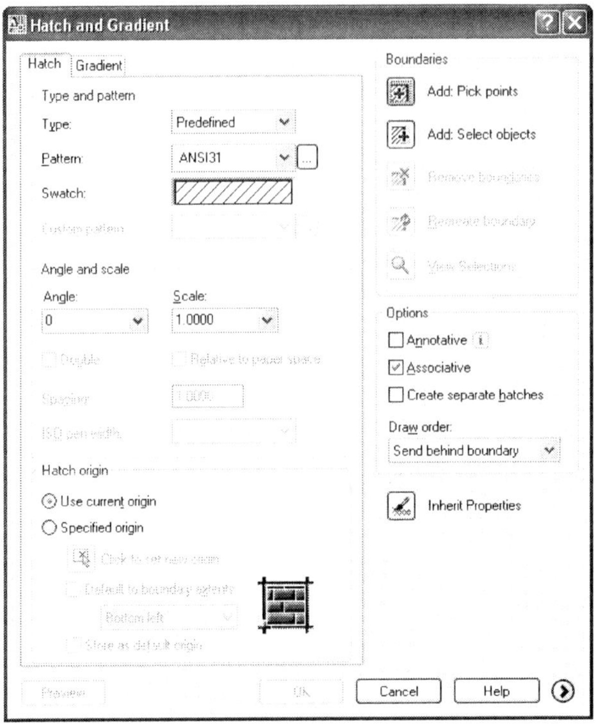

FIGURE 7-1

- Under the Hatch tab, **select** the Type pop-up list. You will get the following choices:
  - User defined.
  - Predefined.
  - Custom.

User Defined
- The most simple hatch pattern is parallel lines. Once this option is selected the following parameters will be valid:
  - **Swatch**, to specify the color of the hatch. We prefer to keep it BYLAYER, as discussed previously.
  - **Angle**, the angle of the parallel lines.
  - **Spacing**, the distance between two parallel lines.
  - **Double**, which means in both ways.
- The following is a **User defined** hatch using **Angle = 45** and **Double** hatch.

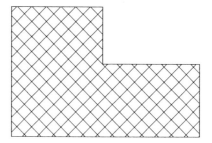

Predefined

There are a handfull of hatch patterns that are availabe for use. You will see ANSI hatches, ISO hatches, and commonly used hatches. The following are the parameters to control:

- **Pattern**; to select the desired pattern either using the pop-up list, the small button with the three dots, or clicking the **Swatch**. The following dialog box will appear showing the **ANSI** hatches:

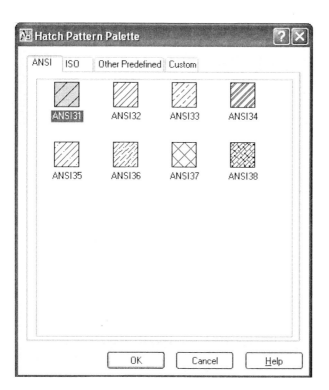

FIGURE 7-2

- Or, click the **ISO** tab and you will see:

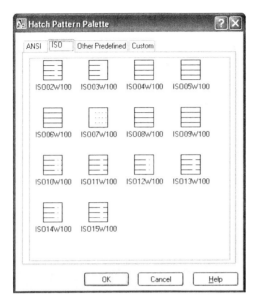

FIGURE 7-3

- Or, click the **Other Predefined** tab and you will see:

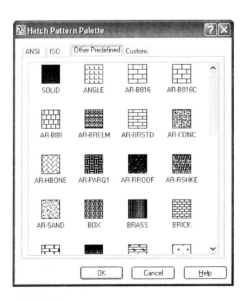

FIGURE 7-4

HATCHING   161

- After you are done selecting the Hatch pattern, click **OK**.
  - Set up the **Angle**.
  - Set up the Scale of the hatch pattern.
  - If you select the ISO hatches, set up the **ISO pen width**.

Custom
- If you customized some hatch patterns, or if you bought some hatch patterns off the Internet, they should be used using **Custom**.

BHATCH COMMAND: SELECTING THE AREA TO BE HATCHED

- After you pick up your hatch pattern and you set the relative parameters, you need to select the desired area to be hatched.
- You can select more than one area in the same command, but the hatch will be considered either as one object, or separate objects.
- There are two ways to select the objects forming the area:
  - Add Pick points.
  - Add Select objects.

Add Pick Points
- This method is very simple. Click inside the desired area. AutoCAD will recognize the area automatically.
- This method will also detect any objects (called islands) within the outer area, and automatically select them to not be hatched.
- Islands can be any object type: circle, closed shape, text, etc.
- We will discuss island detection in detail shortly.

- Click on the button labeled **Add Pick points** located at the top right part of the dialog box.
- The dialog box will disappear temporarily and the following prompt will appear:

```
Pick internal point or [Select objects/remove Boundaries]:
```

- Click inside the desired area(s). Once you are done press [Enter] or right-click, then select the **Enter** option. The dialog box will reappear.

Add Select Objects
- This is the same method to select any object as discussed before.
- Click on the button labeled **Select Objects** located at the top right part of the dialog box. The dialog box will disappear temporarily and the following prompt will appear:

```
Select objects or [picK internal point/remove Boundaries]:
```

- Select the desired objects that form a closed area. Once you are done press [Enter] or right-click, then select the **Enter** option. The dialog box will re-appear.
- After you select, you will be able to do two things:
  - **Remove boundaries** from the selection set. Most likely you will need this button in case of **Pick Points**, because **Pick Points** will select all the inner objects and text as islands. This button will allow you to remove some of the selected objects as islands.
  - **View Selections**; to view the selection set to make sure that it is the right one.

BHATCH COMMAND: PREVIEW THE HATCH

- The data input is complete. You selected the hatch pattern along with its parameters, and you selected the area to be hatched. The next step is to preview the hatch before you make your final decision of accepting the outcome.

- Click the **Preview** button, located at the lower left corner of the dialog box.
- The dialog box will disappear temporarily.
- You will see the results of your settings (i.e., hatch pattern, angle, scale, islands, etc.). Accordingly, AutoCAD will prompt you:

```
Pick or press Esc to return to dialog or <Right-
click to accept hatch>:
```

- Which means, if you like what you see, right-click or press [Enter], otherwise, press [Esc].
- If you pressed [Esc], the dialog will reappear, hence you can change any of the parameters you don't think fit, and you can preview again, and so on.

HATCHING USING BHATCH COMMAND (METRIC)

Workshop 8-A

1. Start AutoCAD 2008, and open the file: **Small_Villa_Ground_Floor_Plan_Metric_Workshop_8. dwg**. (If you solved the previous workshop correctly, open your file.)

2. Make the **Hatch** layer current.
3. Start the Bhatch command and select the **Type** to be **Predefined**.
4. Click on the **Swatch**, select the **Other Predefined** tab, and select the **AR-CONC** pattern.
5. Set the Scale = **100**.
6. Click on the **Pick Points** button and click inside the area representing the outer wall. Press [Enter].
7. Click **Preview** to see the results of the hatching and then press [Enter] to end the command.
8. Start the Bhatch command again and select the **Type** to be **Predefined**.
9. Click on the **Swatch**, select the **ANSI** tab, and select the **ANSI32** pattern.
10. Set the Scale = **500**.
11. Click on the **Pick Points** button and click inside the area representing the inner walls. Press [Enter].
12. Click **Preview** to see the results of the hatching and then press [Enter] to end the command.
13. Save and close the file.

HATCHING USING BHATCH COMMAND (IMPERIAL)

 Workshop 8-B

1. Start AutoCAD 2008, and open the file: **Small_Villa_Ground_Floor_Plan_Imperial_Workshop_8. dwg**. (If you solved the previous workshop correctly, open your file.)
2. Make the **Hatch** layer current.
3. Start the Bhatch command and select the **Type** to be **Predefined**.
4 Click on the **Swatch**, select the **Other Predefined** tab, and select the **AR-CONC** pattern.
5. Set the Scale = **5**.
6. Click on the **Pick Points** button and click inside the area representing the outer wall. Press [Enter].
7. Click **Preview** to see the results of the hatching and then press [Enter] to end the command.
8. Start the Bhatch command again and select the **Type** to be **Predefined**.
9. Click on the **Swatch**, select the **ANSI** tab, and select the **ANSI32** pattern.

10. Set the Scale = **20**.
11. Click on the **Pick Points** button and click inside the area representing the inner walls. Press [Enter].
12. Click **Preview** to see the results of the hatching and then press [Enter] to end the command.
13. Save and close the file.

BHATCH COMMAND: OPTIONS

- While you are hatching, there are some options in Bhatch command that you should know in order to have full control over the process of hatching.
- On the right side of the dialog box there is a part labeled **Options**:

Annotative
- This is an advanced feature, which will be discussed in other levels beyond this book.

Associative
- Associative means that there is a relationship between the hatch and the boundary—whenever the boundary changes the hatch changes automatically.
- Keeping this option on is good for the user.

Example
- Assume you have the following shape to hatch:

- You started the Bhatch command, made sure the **Associative** checkbox is on, and hatched the shape. This will be the outcome:

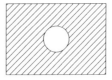

- Now, select the circle, move it to right, and see how hatch will react:

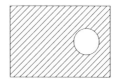

- As you can see, the hatch reacted properly to the movement of the circle, which proves that the boundary and the hatch are **Associative**.
- Erase the hatch, then do the same procedure. This time make sure that **Associative** is off. This is the result of the movement of the circle:

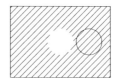

- You can see from above, hatch didn't react properly to the movement of the circle as a result of not turning the **Associative** checkbox on.

Create Separate Hatches
- In previous AutoCAD releases, when you hatched several areas using the same command, all of the hatches were considered one entity, hence they moved togather and erased togather.
- Now you can hatch several areas using the same command and each hactch will be considered a separate entity. To do so, check the **Create separate hacthes** checkbox on.

Draw Order
- In cases of hatched areas intersecting with other hatched areas (especially hatched with Solid hatching) you need to set the Draw order while you are inserting the hatch, so you can ensure the right appearance of each area:
- The four cases are:
  - Send to back
  - Bring to front (see the following example)

Bring to Front   Send to Back

Before   **After**

- Send behind boundary

- Bring in front of boundary (see the following example)

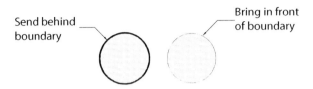

| Inherit Properties | - Below the **Options** part there is a button that is labeled **Inherit Properties**. |

- The purpose of this button is to help the user quickly hatch a new area with the same exact features of an existing hatch.

- Click the **Inherit Properties** button. The dialog box will disappear temporarily and the following prompt will appear:

```
Select hatch object:
```

- The mouse will change to a painting brush. Click the source hatch pattern.
- The following will happen:
  - A new prompt will appear:

```
Inherited Properties: Name <SOLID>, Scale <1.0000>
Angle <0>
Pick internal point or [Select objects/remove
Boundaries]:
```

- Meanwhile the mouse cursor will change to a bigger paint brush with small crosshairs, so you can click inside the new areas you want to hatch.
- Click inside the desired area(s). Once you are done press [Enter] or right-click, then select the **Enter** option. The dialog box will reappear. Press **OK** to finish the command.

BHATCH COMMAND: HATCH ORIGIN

- Each area has something called a "Center of Area."
- AutoCAD used to always use this concept while creating the hatching. So if you hatched a circle, AutoCAD would start from the center of the area and go to all directions from there, filling the whole area.
- AutoCAD now gives the user the option to specify a new origin other than the default one.

- At the left lower side of the dialog box you will see a part labeled **Hatch origin**:

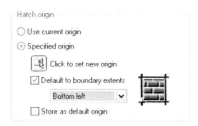

- Click the **Specified origin** radio button.
- In order to specify a new origin you have two options:

  - Use the button labeled **Click to set new origin** and select any point your desire.
  - Or click on the checkbox labeled **Default to boundary extents** and select one of the options available: Bottom left, Bottom right, Top left, Top right, or Center. You can also set this choice as the default origin always.
- See the below example:

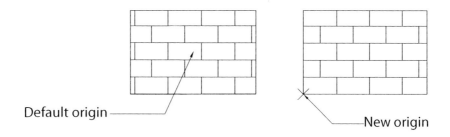

Default origin

New origin

ASSOCIATIVE HATCHING AND HATCH ORIGIN (METRIC)

Workshop 9-A

1. Start AutoCAD 2008, and open the file: **Small_Villa_Ground_Floor_Plan_Metric_Workshop_9. dwg**. (If you solved the previous workshop correctly, open your file.)
2. Make the **Hatch** layer current.
3. Zoom to the kitchen at the lower right of the plan.

4. Start Bhatch command and select the **Type** to be **Predefined**.
5. Click on the **Swatch**, select the **Other Predefined** tab, and select the **ANGLE** pattern.
6. Set the Scale = **1000** and make sure that Hatch origin = **Use current origin**. Also, make sure that **Associative** is on.
7. Click on the **Pick Points** button and click inside the area representing the bathroom. Press [Enter].
8. Click **Preview** to see the results of the hatching. Press [Esc] and you will see the dialog box again. Select **Specified origin** and specify the lower left corner of the kitchen to be the new origin. Click **Preview** again, do you see any change? Press [Enter] to finish the command.
9. Move the oven to the right, what is happening to the hatch? Does it react correctly or not?
10. Save the file and close it.

ASSOCIATIVE HATCHING AND HATCH ORIGIN (IMPERIAL)

Workshop 9-B

1. Start AutoCAD 2008, and open the file: **Small_Villa_Ground_Floor_Plan_Imperial_Workshop_9. dwg**. (If you solved the previous workshop correctly, open your file.)
2. Make the **Hatch** layer current.
3. Zoom to the kitchen at the lower right of the plan.
4. Start the Bhatch command and select the **Type** to be **Predefined**.
5. Click on the **Swatch**, select the **Other Predefined** tab, and select the **ANGLE** pattern.
6. Set the Scale = **50** and make sure that Hatch origin = **Use current origin**. Also, make sure that **Associative** is on.
7. Click on the **Pick Points** button and click inside the area representing the bathroom. Press [Enter].
8. Click **Preview** to see the results of the hatching. Press [Esc] and you will see the dialog box again. Select **Specified origin** and specify the lower left corner of the kitchen to be the new origin. Click **Preview** again, do you see any change? Press [Enter] to finish the command.
9. Move the oven to the right, what is happening to the hatch. Does it react correctly or not?
10. Save the file and close it.

BHATCH COMMAND: ADVANCED FEATURES

- At the lower right corner of the dialog box you will see a small button with an arrow pointing to the right. Click this button and you will see more advanced options for hatching:

Islands
- The first part is **Islands**:

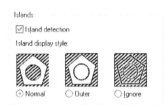

- Islands in AutoCAD are the inner objects inside the outer boundary of an area to be hatched.
    - Click **Islands detection** off if you don't want AutoCAD to recognize the inner objects.
    - Click **Islands detection** on if you want AutoCAD to select the inner objects as boundaries, and hence not to hatch them.
- Select one of these three styles:
    - **Normal** means in cases of three objects or more inside each other AutoCAD will hatch the outer, leave the second, hatch the third, etc.
    - **Outer** means in cases of three objects or more inside each other AutoCAD will hatch the outer only, leaving the inner objects intact.
    - **Ignore** means in cases of three objects or more inside each other AutoCAD will ignore all the inner objects and hatch the outer area fully.

Boundary retention
- By default, AutoCAD creates a polyline around the detected area and once the hatch command is finished, AutoCAD will delete this polyline.
- In this part you can keep this temporary polyline.

- Click the **Retain boundaries** checkbox on, then specify the **Object type** (the other object type is Region).

Boundary Set
- When a user is using the **Add: Pick points** option to define the boundary to be hatched, by default AutoCAD will analyze all the objects in the current viewport.

- This may take a very long time depending on the complexity of the drawing. To minimize the time you can provide a selection set for AutoCAD to analyze the boundary from.
- The **Boundary set** part of the dialog box looks like:

- By default the selected option is **Current viewport**.
- Click the button labeled **New** and the dialog box will disappear temporarily. The following prompt will appear:

```
Select objects:
```

- Select the desired objects and press [Enter] or right-click. The dialog box will appear again, but this time the selected option will be **Existing set**, as follows:

- Now when you ask AutoCAD to select a boundary by clicking inside it, AutoCAD will not analyze all objects in the current viewport, but rather analyze only the objects you selected.

Gap Tolerance

- AutoCAD used to ask for a closed area to make the hatching process successful, but now AutoCAD adds some flexibility as it accepts to hatch an area with a small gap. How small is up to the user.
- You can set the maximum gap that AutoCAD may ignore.

Inherit Options

- This part deals with two things we discussed in the previous pages (i.e., Inherit Properties and Hatch origin).
- When you Inherit Properties from an existing hatch, which Hatch origin should be used?
  - Current origin.
  - Source hatch origin.

HATCHING USING TOOL PALETTES

- Previously we introduced the concept of Tool Palettes, which includes both blocks and hatches.
- The main feature of Tool Palettes is the drag-and-drop feature. So we will utilize this feature to speed up the process of hatching
- Do the following steps:
  - Create a new Tool Palette (call it My Hatches).
  - Use Bhatch to add hatches to your different drawings.
  - While you are hatching you are changing the Hatch settings.
  - Each and every time you feel like this hatch, with all of its settings, will be used in other drawings, simply Drag-and-Drop it in your newly created Tool Palette.
  - You can make several copies of your hatch. Then you can use customization to change the different settings of each hatch. Right-click on any hatch in your Tool Palette and select **Properties**, the following dialog box will appear:

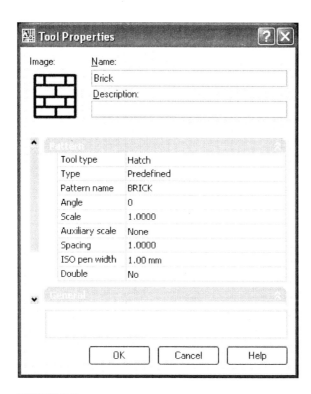

FIGURE 7-5

- After several drawings, you will have a big library of hatches that you will use frequently in your drawings.
- Now use the Drag-and-Drop from your Tool Palette to your drawing.

- By default there will be a Tool Palette (called **Hatches**) that you can utilize if you don't want to create your own.

HATCHING AND TOOL PALETTE (METRIC AND IMPERIAL)

Workshop 10-A & 10-B

1. Start AutoCAD 2008, and open the file: **Small_Villa_Ground_ Floor_Plan_Metric_Workshop_10. dwg**. (If you solved the previous workshop correctly, open your file.) Or, open the file: **Small_Villa_ Ground_Floor_Plan_Imperial_Workshop_10. dwg**. (If you solved the previous workshop correctly, open your file.)
2. Start Tool Palettes, create a new Tool Palette, and name it **My Hatches**.
3. Drag and drop the three hatches we used in our file, which are: AR-CONC, ANSI32, and ANGLE.
4. In the new Tool Palette select any of the three hacthes, right-click, and select **Properties**. Make sure that the Layer is always **Hatch** and not **use current**.
5. Next time you use the hatch from the Tool Palette you won't worry "in which layer the hatch will reside."
6. Save the file and close it

GRADIENT COMMAND

- Use the **Gradient** command if you to want to shade a 2D area with one color along with white, black, or something in between, or with two colors.
- It uses the same method of Bhatch command, so no need to reiterate it again.
- Use one of the following methods to issue the command:
  - From the dashboard, and using the **2D Draw** toolbar and the small triangle at the right, click and hold, then select the **Gradient** button.
  - From menus select **Draw/Gradient**.
  - Type **gradient** in the Command window.

- The following dialog box will appear:

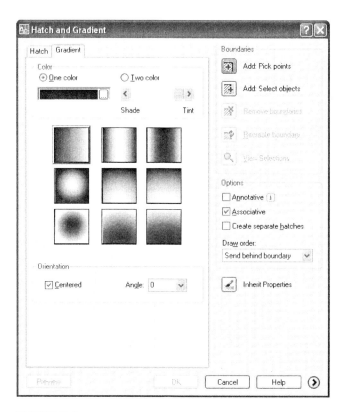

FIGURE 7-6

- You start by specifying whether you want to use either one color, along with the color white, or you want to use two colors.

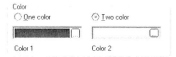

- If you select **One color**, then control two things:
  - Click the small button to select the desired color.
  - Control the slider from **Tint** (total white) to **Shade** (total black), or any color between them.

- If you select **Two color**, the following will appear:

How to Select a Color

- There are three sets of colors you can select from:
  - 255 color called Index Color:

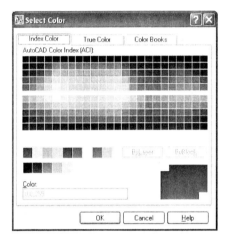

FIGURE 7-7

- 24-bit True color. You can select from two models HLS (Hue, Luminance, and Saturation):

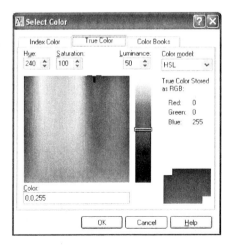

FIGURE 7-8

- Or RGB (Red, Green, and Blue):

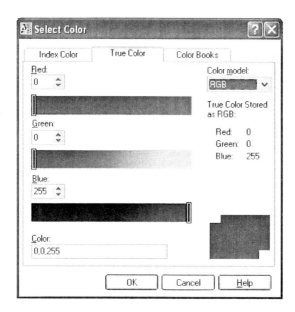

FIGURE 7-9

- Or finally, select a color from one of 11 color books available:

FIGURE 7-10

- Now select one of the nine gradient patterns

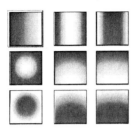

- Select whether your pattern is symmetrical (Center) or not. And the angle of the pattern:

- The rest is identical to the Bhatch command.

HOW TO EDIT AN EXISTING HATCH

- Use the **Hatchedit** command if you want to edit an existing hatch (whether using the Bhatch command, Drag-and-Drop, or the Gradient command).
- There are several ways to issue this command:
  - From the **Modify** II toolbar, click the **Edit Hatch** button.
  - From menus **select Modify/Object/Hatch**.
  - Type **hatchedit** (or **he**) in the Command window.
  - Select the hatch, right-click, and from the shortcut menu select the **Hatch Edit** option.
  - Double-click the hatch you want to edit.
- Regardless of the method used, AutoCAD will show the following prompt:

```
Select hatch object:
```

- Select the desired hatch and the Bhatch dialog box will appear. Change the settings as you wish, then click **OK**.
- Another way to edit a hatch is:
  - Select the hatch.
  - Right-click and select **Properties** from the shortcut menu.

- The following **Properties** palette will appear:

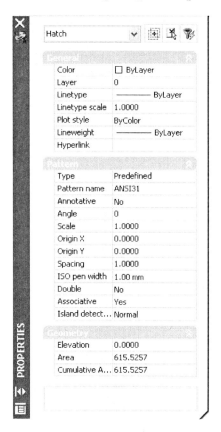

FIGURE 7-11

- In the **Properties** palette you can edit all the data related to the hatch selected (some of the settings will be applicable to User hatch patterns like Spacing and Double).

 Recreate Boundary
- If you hatched an area, and for one reason or another the boundary was lost and the hatch was kept, you can use this option to recreate the boundary of an existing hatch.
- Use any method to edit the hatch without boundary. The hatch dialog box will appear. Do the following steps:

  - Click the **Recreate boundary** button. The dialog will disappear temporarily and the following prompt will appear:

```
Enter type of boundary object [Region/Polyline]
<Polyline>:
```

- Type **P** for polyline or **R** for Region and the following prompt will appear:

  ```
  Associate hatch with new boundary? [Yes/No] <Y>:
  ```

- Type **Y** for yes or **N** for no.
- The dialog box will appear again. Click **OK** to finish the command.

EDIT HATCHING (METRIC)

Workshop 11-A

1. Start AutoCAD 2008, and open the file: **Small_Villa_Ground_Floor_Plan_Metric_Workshop_11. dwg**. (If you solved the previous workshop correctly, open your file.)
2. Select the AR-CONC hatch, right-click, then select **Hatch Edit**. The dialog box of Hatch Edit will appear. Change the **Scale** of the hatch to **75**.
3. Click **Preview**, then press [Enter] to accept the changes you made.
4. Double-click the ANGLE hatch and change the Angle to 45.
5. Click **Preview**, then press [Enter] to accept the changes you made.
6. Save and Close.

EDIT HATCHING (IMPERIAL)

Workshop 11-B

1. Start AutoCAD 2008, and open the file: **Small_Villa_Ground_Floor_Plan_Imperial_Workshop_11. dwg**. (If you solved the previous workshop correctly, open your file.)
2. Select the AR-CONC hatch, right-click, then select **Hatch Edit**. The dialog box for Hatch Edit will appear. Change the **Scale** of the hatch to **2.5**.
3. Click **Preview**, then press [Enter] to accept the changes you made.
4. Double-click the ANGLE hatch and change the Angle to 45.
5. Click **Preview**, then press [Enter] to accept the changes you made.
6. Save the file and close it.

CHAPTER REVIEW

1. The origin of a hatch area is defined by AutoCAD and you can't change it.
   a. True
   b. False
2. You can create a boundary for an existing hatch using the _____ button.
3. AutoCAD now supports 24-bit true colors.
   a. True
   b. False
4. Which of these statements is NOT true about hatching in AutoCAD:
   a. You can set the Draw order of the Hatch.
   b. You can use the conventional method and drag-and-drop method.
   c. You have to have a closed area to hatch it.
   d. You can use Hatch or Gradient colors.
5. One of these, Bhatch command can't do:
   a. Separate hatches using the same command.
   b. Hatching areas with a gap.
   c. Set the scale of the hatch pattern.
   d. Hatch with three color gradients.
6. If you want the hatch to react to any change in the boundary, click the _____ checkbox in the Bhatch dialog box.

CHAPTER REVIEW ANSWERS

1. b
2. Recreate boundary
3. a
4. c
5. d
6. Associative

Chapter 8
Text and Tables

In This Chapter
- How to create Text styles
- How to write using the old method of text: DTEXT
- How to write using the new method of text: MTEXT
- Editing Text in AutoCAD (contents and properties)
- Spelling check, Find and Replace
- How to create Table styles
- How to insert Tables

INTRODUCTION

- In order to write text in AutoCAD you first need to create your own **Text Style**.
- In Text Style you specify the characteristics of your text which will apply to all the texts you write in your file.
- You should have a good number of Text Styles in your drawing in order to cover all the requirements (big fonts for titles, small fonts for remarks, special text style for dimensions, etc.)
- Text Style can be shared between files using Design Center.
- After you create your text style, you can use several commands to write text in you drawing:
  - Single line text (old method).
  - Multiline text (new method).
- After you finish writing you can edit and spell check the written text.
- In order to create Tables with text, you have to create a Table Style.
- Then you can insert tables and write text inside them.
- Table styles can be shared between files using Design Center.

TEXT STYLE

- The first step in writing text in AutoCAD is to create a text style.
- Text style is where you define the characteristics of your text.
- To issue the command use one of the following ways:
    - From the dashboard and using **Text** panel, click the **Text Style** button.
    - From menus select **Format/Text Style**.
    - Type **style** (or **st**) in the Command window.
- The following dialog box will appear:

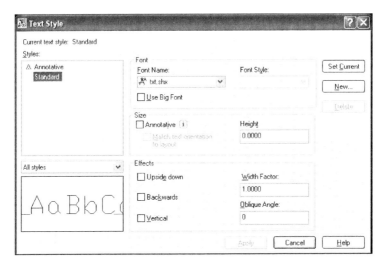

FIGURE 8-1

- As you can see, AutoCAD comes with a default text style called **Standard**.
- This style contains the default settings for the text style.
- This style is very simple and users should consider creating their own.
- To create a new text style, click the **New** button. The following dialog box will appear:

FIGURE 8-2

- Type in the name of the new text style (use the same naming convention as the layers).
- When done, click **OK**.

Font Name
- The fisrt thing to do is to select the desired font.
- There are two types of fonts that you can use in AutoCAD:
  - Shape files (*.shx), the old method for fonts.
  - True Type fonts (*.ttf), the new method for fonts.
- See the following illustration to differentiate between the two types:

True Type fonts   Shape files fonts

Font Style
- If you select a True Type font, you will be able to select the Font Style. You have the following choices to pick from:
  - Regular
  - Bold
  - Bold Italic
  - Italic
- See the following illustration:

Regular   Bold   Bold/Italic   Italic

Annotative
- This is an advanced feature, keep it off for now.

Height
- Specify the height of the text, see the following illustration:

- As you can see from the illustration, the Height mentioned in the dialog box is for the capital letters. Automatically 2/3 of the height will be for lowercase letters, and 1/3 will be for letters that descend below the baseline.
- There are two methods for specifying the height of text:
  - Leave the value equal to 0 (zero), which means you will have to specify the height each and every time you use this style.
  - Specify a height value that will always be used once you use this style.

Effects
- There are five effects you can add to your text:
  - **Upside down** effect, look at the following illustration:

Bag — Normal
Bag (upside down) — Upside down

  - **Backward** effect, is to write from right to left:

Bag — Normal
Bag (mirrored) — Backward

  - **Width Factor** effect, is the Width/Height:

Bag — Width Factor = 1
Bag — Width Factor = 1.5
Bag — Width Factor = 0.75

  - **Oblique Angle** effect is as follows:

Bag — Oblique Angle = 0
Bag — Oblique Angle = +15
Bag — Oblique Angle = -15

- **Vertical** effect is only applicable for .shx fonts, and it will write the text from top to bottom.
 - Whenever you are done, click the **Apply** button, then the **Close** button.
- At the left part of the dialog box there is a pop-up list showing **All styles**. Using this list you can show in this dialog box all defined text styles whether used or not, or only the text styles that are used in this drawing.

CREATING A TEXT STYLE (METRIC)

 Workshop 12-A

1. Start AutoCAD 2008, and open the file: **Small_Villa_Ground_Floor_Plan_Metric_Workshop_12.dwg**. (If you solved the previous workshop correctly, open your file.)
2. Create a text style named = **Title** with the following settings:
   a. Font = **Arial**
   b. Font Style = **Bold**
   c. Height = **900**
   d. Width Factor = **2**
3. Create a text style named = **Inside_Annot** with the following settings:
   a. Font = **Times New Roman**
   b. Font Style = **Regular**
   c. Height = **300**
   d. Width Factor = **1.0**
4. Create a text style named = **Dimension** with the following settings:
   a. Font = **Arial**
   b. Font Style = **Regular**
   c. Height = **400**
   d. Width Factor = **1.0**
5. Save the file and close it.

CREATING A TEXT STYLE (IMPERIAL)

 Workshop 12-B

1. Start AutoCAD 2008, and open the file: **Small_Villa_Ground_Floor_Plan_Imperial_Workshop_12.dwg**. (If you solved the previous workshop correctly, open your file.)

2. Create a text style named = **Title** with the following settings:
   a. Font = **Arial**
   b. Font Style = **Bold**
   c. Height = **3′–0″**
   d. Width Factor = **2**
3. Create a text style named = **Inside_Annot** with the following settings:
   a. Font = **Times New Roman**
   b. Font Style = **Regular**
   c. Height = **1′–0″**
   d. Width Factor = **1**
4. Create a text style named = **Dimension** with the following settings:
   a. Font = **Arial**
   b. Font Style = **Regular**
   c. Height = **1′–4″**
   d. Width Factor = **1**
5. Save the file and close it.

DTEXT COMMAND

- The DTEXT command is the first of two commands you can use in order to write text in AutoCAD.
- This command is also called Single line text.
- Although you write several lines of text in each command, each would be considered as a separate object.
- To issue this command use one of the following methods:
  - From the dashboard and using the **Text** panel, click the **Single Line Text** button.
  - From menus select **Draw/Text/Single Line Text**.
  - Type **dtext** (or **dt**) in the Command window.
- Either way, the following prompt will appear:

```
Current text style: "arial_09" Text height: 0.9000
Annotative No
Specify start point of text or [Justify/Style]:
```
(Specify the start point of the baseline)
```
Specify rotation angle of text <0>:
``` 
(Specify the rotation angle of the baseline)

- As noted above, AutoCAD will use the current Text Style to write the desired text. In order to change it, use the **Text** panel as explained below:

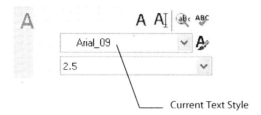

Current Text Style

- After you input these values, start writing. If you want a new line, press [Enter]. The cursor will go to the beginning of a new line. Continue doing the same. Once you are done press [Enter] twice.
- Below is an example of 45-degree, baseline single-line text:

MTEXT COMMAND

- The MTEXT command was first introduced in AutoCAD R13, and continues improving since then.
- It simulates MS Word simplicity in creating text, hence it is easier and handier for normal users that possess experience with MS Word.
- This command is also called Multiline text.
- All the text you write in a single command would be considered a single object.
- To issue this command use one of the following methods:
    - From the dashboard and using the **Text** panel, click the **Multiline Text** button.
    - From menus select **Draw/Text/Multiline Text**.
    - Type **mtext** (or **mt**) in the Command window.
- Either way, AutoCAD will show the following prompt:

```
Current text style: "arial_09" Text height: 0.9000
Annotative No
```

```
Specify first corner: (Specify first corner)
Specify opposite corner or [Height/Justify/Line
spacing/Rotation/Style/Width]: (Specify opposite
corner)
```

- AutoCAD wants users to select two opposite corners to specify the area that you will write in. Check the illustration below:

- After you specify the two corners, the **In-Place text editor** with a ruler will appear along with a **Text Formatting** toolbar. Check the picture below:

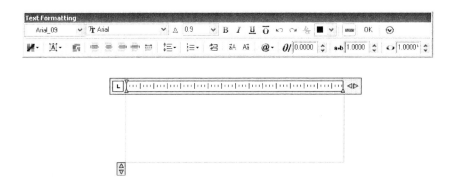

- A blinking cursor will appear in the **In-Place text editor** so you can type your desired text. Using the **Text Formatting** toolbar, you can format the text as you wish.
- If you created a text style (highly recommended), then you will see at the upper right part of the toolbar the name of the current text style along with the name of the font and the height.
- In order to format a text you should select it first—just like you do in MS Word.
- You can format the text to be **Bold**, **Italic**, **Underlined**, and **Overlined**.
- If you make a mistake you can use **Undo** and **Redo**.

■ If you have a fraction and you want it to appear stacked, click on the **Stack** button (of course you have to select the fraction first). Check the example below:

$$3/4 \qquad \frac{3}{4}$$

Not Stacked   Stacked

■ You can also change the color of the text (to become not BYLAYER).
■ You can also Show or Hide the ruler.
 ■ Mtext Justification, to allow the user to set the justification for the text, is as follows:
 • **Top Left (TL)**, the text will be vertical at the top of the area specified and horizontal to the left of the area.
 • **Top Center (TC)**, the text will be vertical at the top of the area specified and horizontal at the center of the area.
 • **Top Right (TR)**, the text will be vertical at the top of the area specified and horizontal to the right of the area.
 • **Middle Left (ML)**, the text will be vertical at the middle of the area specified and horizontal to the left of the area.
 • **Middle Center (MC)**, the text will be vertical at the middle of the area specified and horizontal at the center of the area.
 • **Middle Right (MR)**, the text will be vertical at the middle of the area specified and horizontal to the right of the area.
 • **Bottom Left (BL)**, the text will be vertical at the bottom of the area specified and horizontal to the left of the area.
 • **Bottom Center (BC)**, the text will be vertical at the bottom of the area specified and horizontal at the center of the area.
 • **Bottom Right (BR)**, the text will be vertical at the bottom of the area specified and horizontal to the right of the area.
■ You can set the horizontal **Justification** of the text. Select whether to Left, Center, Right, Justify, or Distribute the area the text.
■ You can set the Line Spacing of the paragraph. You have the choice of 1.0x, 1.5x, 2.0x, 2.5x, or you can set your own.
■ If you want to use **Bullets and Numbering** you have 3 choices, using letters (lowercase or uppercase), numbers, or bullets.
■ You can change the case of the text from lowercase to uppercase and vise versa.

- You can add Symbols to your text. If you click the **Symbol** button the following menu will appear. You can select to add one of 20 available symbols:

Degrees	%%d
Plus/Minus	%%p
Diameter	%%c
Almost Equal	\U+2248
Angle	\U+2220
Boundary Line	\U+E100
Center Line	\U+2104
Delta	\U+0394
Electrical Phase	\U+0278
Flow Line	\U+E101
Identity	\U+2261
Initial Length	\U+E200
Monument Line	\U+E102
Not Equal	\U+2260
Ohm	\U+2126
Omega	\U+03A9
Property Line	\U+214A
Subscript 2	\U+2082
Squared	\U+00B2
Cubed	\U+00B3
Non-breaking Space	Ctrl+Shift+Space
Other...	

- Just like Oblique angle discussed in the text style section.

- To increase or decrease the spaces between letters. Values greater than 1 means more spaces and vise versa.

- Just like the Width factor discussed in the text style section.

- When you are done writing and formatting, click **OK** to save and finish the command.
- There are other things you can do while you are in the In-Place editor, we will mention two of these:
  - Import Text.
  - Set the Indents.

Import Text

- If you created a nonformatted text (*.txt) using any editor like Notepad, you can bring the text to the In-Place editor and then format it.
- While you are in the In-Place editor, right-click and a long shortcut menu will appear. Select **Import Text** and the following dialog box will appear:

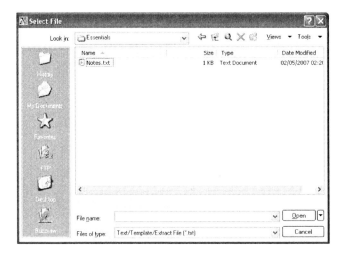

FIGURE 8-3

- Select the desired drive and folder, then select the desired txt file. Once you click **Open** you will see the text landing in the In-Place editor for formatting.

Set Indents
- Setting Indents in the In-Place editor is identical to the process in MS Word. See the illustration below:

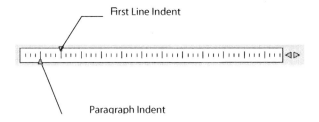

- Move the **First Line Indent** to specify where the first line will start.
- Move the **Paragraph Indent** to specify where the next line will start.
- Also, you can set **Tabs** for your text by clicking anywhere in the ruler.

WRITING TEXT (METRIC AND IMPERIAL)

 Workshop 13-A & 13-B

1. Start AutoCAD 2008, and open the file: **Small_Villa_Ground_Floor_Plan_Metric_Workshop_13.dwg**. (If you solved the previous

workshop correctly, open your file.) Or, open the file: **Small_Villa_ Ground_Floor_Plan_Imperial_Workshop_13.dwg**, (If you solved the previous workshop correctly, open your file.)

2. Look at the picture below:

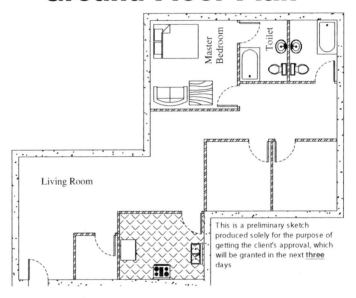

3. Make the **Text** layer the current layer.
4. Using Single Line Text (DTEXT command) and using Text Style = **Inside_Annot**, type the following words: **Master Bedroom**, **Toilet**, **Living Room**.
5. Using Single Line Text (DTEXT command) and using Text Style = **Title**, type the following words: **Ground Floor Plan**.
6. Using the Multiline Text command, specify the area in the lower right part of the plan as shown above, and using the Text style = Inside_Annot, type the following text (with the mistakes in them): *"This is a preliminary sketch produced solely for the purpose of getting the client's approval which will be granted in the next three days"* Don't close the editor.
7. Select the word three and make it red, bold, and underlined. Click OK to close the editor.
8. Save the file and close it.

EDITING TEXT

- In order to edit the *contents* of the text, simply double-click the text.
- If you double-click multiline text, the In-Place editor will appear with the **Formatting Text** toolbar for further adding/deleting or simply reformatting.
- If you double-click single-line text, the text will be available for adding and deleting.
- Other ways are:

  - From the **Text** toolbar, click the **Edit** button.
  - Type **ddedit** in the Command prompt.
- This command can go with both types of texts.
- Also, you can select multiline text, right-click, and select **Mtext Edit**, which is equivalent to the **mtedit** command as a typed version of the command.

EDITING TEXT PROPERTIES

Single-Line Text

In order to edit the properties of single-line text, simply click on it, then right-click. A shortcut menu will appear. Select **Properties** and the following palette will appear:

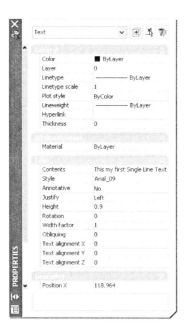

FIGURE 8-4

- You can change the **General** properties of single-line text (Color, Layer, Linetype, etc.).
- You can change the Contents of the text and other properties such as Style, Justification, Height, Rotation, etc.
- You can change the Geometry of the text (position of X, Y, and Z).
- Finally, you can change the **Miscellaneous** properties of single-line text such as Upside down and Backward.

Multiline Text
- To edit the properties of multiline text, simply click on it, then right-click. A short-cut menu will appear. Select **Properties** and the following palette will appear:

FIGURE 8-5

- You can change the General properties of multiline text (Color, Layer, Linetype, etc.).
- You can change the contents of mutliline text and the other properties such as Style, Justification, Direction, Height, Rotation, etc.
- You can change the Geometry of the text (position of X, Y, and Z).
- If you select single-line text and multiline text, you change only the **General** properties.
- You can select either multiple single-line text, or multiple multiline text, and change their properties in one shot.

TEXT AND GRIPS

- If you click (single-click) over single-line text you will see the following:

Elevation

- The grip appears at the start point of the Baseline.
- On the other hand, if you click on multiline text you will get the following:

The Elevation and the cross-section views will be modified together

- The grips appear at the four edges of the multiline text. This is very helpful if you want to change the area the multiline text is occupying.
- You can make it wider by making one of the righthand grips hot and dragging it to the right side, like this:

The Elevation and the cross-section views will be modified together

- Or you can make it narrower by making one of the righthand grips hot and dragging it to the left side, like this:

The Elevation and the cross-section views will be modified together

- While you are selecting multiline text, right-click and a shortcut menu will appear. One of the options is **Mtext Edit**, so you will be able to edit the text.

SPELLING CHECK AND FIND AND REPLACE

Spelling Check

- AutoCAD can check the spelling of any text whether Single-line text or Multiline text.
- AutoCAD will spell check the **Entire drawing**, **Current space/layout**, or the **Selected text**.
- Issue the command using one of the following methods:

  - From the dashboard and using the **Text** panel, click the **Spell Check** button.
  - From menus select **Tools/Spelling**.
  - Type **spell** in the Command window.
- The following dialog box will appear:

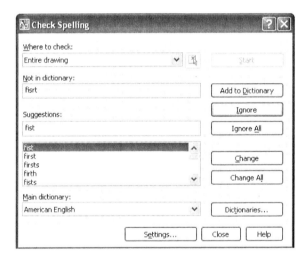

FIGURE 8-6

- This is identical to the MS Word spelling checker.
- If AutoCAD finds any word wrong it will give you suggestions and you can choose one of them. You have the ability to change or ignore.

Find and Replace

- AutoCAD can find any word or part of a word in the entire drawing file and replace it at user's wish.
- Issue the command using one of the following methods:

  - From the dashboard and using the **Text** panel, select the **Find** button.
  - From menus select **Edit/Find**.
  - Type **find** in the Command window.

- Either way, the following dialog box will appear:

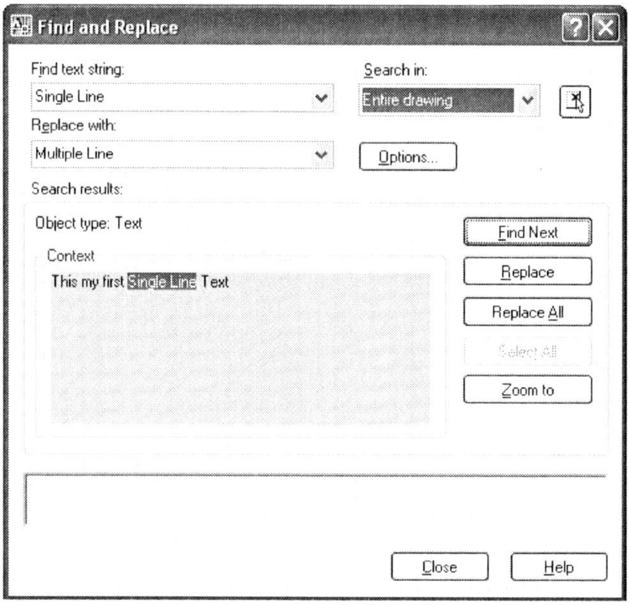

FIGURE 8-7

- Under **Find text string**, type the word(s) you want to find.
- Under **Replace with**, type the new word(s) you want to replace.
- You can search in the **Entire drawing** or use a selection of text to search.
- You have three choices to select from: **Find**, **Replace**, and **Replace All**.
- When you are done, click **Close**.

EDITING TEXT (METRIC AND IMPERIAL)

 Workshop 14-A & 14-B

1. Start AutoCAD 2008, and open the file: **Small_Villa_Ground_ Floor_Plan_Metric_Workshop_14.dwg**. (If you solved the previous workshop correctly, open your file.) Or, open the file: **Small_Villa_ Ground_Floor_Plan_Imperial_Workshop_14.dwg**, (If you solved the previous workshop correctly, open your file.)

2. Select the Mulitline text, the four grips will appear. Select one of the right grips to make it hot, and stretch it to the right so you make the text one line less.
3. Double-click the Multiline text and make the following changes:
   a. Select the word "solely" and make it italic.
   b. Add a comma before the word "which"
   c. Press [Enter] after the last word to add a new line and type your initials.
4. Start the Spelling checker of AutoCAD and select the Multiline text.
5. Correct the wrong words accordingly.
6. Save the file and close it.

TABLE STYLE

- Just like we did with text; to create a good table in AutoCAD, do the following two steps:
  - Create a Table Style.
  - Insert and Fill the table.
- In **Table Style** you will define the main features of your table.
- To issue this command use one of the following methods:

  - From the dashboard and using the **Tables** panel, click the **Table Style** button.
  - From menus select **Format/Table Style**.
  - Type **tablestyle** in the Command window.
- You will see the following dialog box:

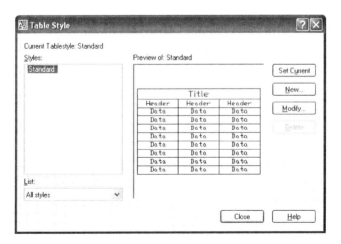

FIGURE 8-8

- As you can see, there is a predefined style called **Standard**.
- There will be a preview (always) that will show you the changes you are making, hence, it will be easy for you to decide right from wrong.
- To create a new table style, click the **New** button. You will see the following dialog box:

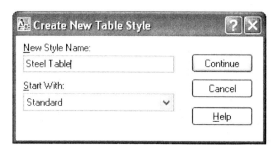

FIGURE 8-9

- Type in the name of your new style.
- Select the **Start With** style (you will start with a copy from this style).
- Click the **Continue** button and the following dialog box will appear:

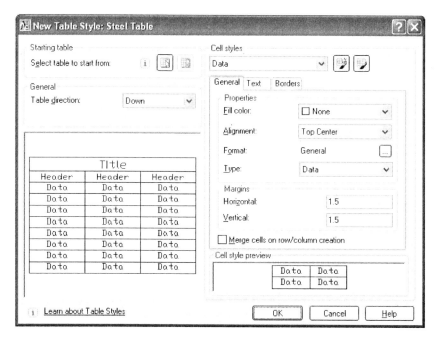

FIGURE 8-10

- Under **Starting table**, the user can select an existing table and copy its style, so as not to start from scratch.
- Under **General**, specify the **Table direction**, whether:
  - **Down**, this means the title and column headers are at the top of the table and the cells will go below them.
  - **Up**, this means the title and column headers are at the bottom of the table and the cells will go above them.
- Under **Cell styles**, you have three choices to pick from: **Data**, **Header**, and **Title**. This is to control the table's three parts. You can control the **General** properties, **Text** properties, and **Border** properties of the three parts.

General Tab

- **General properties** part is:

- Control the following settings:
  - **Fill Color**, select whether the cells will have a colored background or not.
  - **Alignment**, select the justification for the text compared to the cell (you have nine choices to select from). To illustrate the last two points, see the example below:

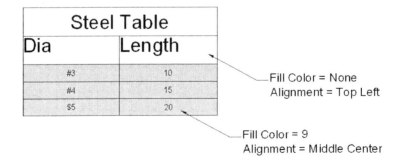

- **Format**, select the format of the numbers and click on the small button with the three dots. You will see the following dialog box:

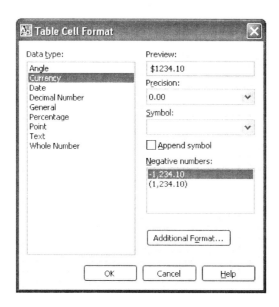

FIGURE 8-11

- Set the **Type** to Data or Label.
- Under **Margins**, control the **Horizontal** and **Verical** distances taken around the Data relative to the borders.

Text Tab • **Text properties** part is:

- Control the following settings:
  - **Text style** you will use in the cells.
  - **Text height** to be used (this is only applicable if the Text style selected has a Height = 0).
  - **Text color** to be used (most likely you should leave it Bylayer or Byblock).
  - **Text angle**, sets the oblique angle of the text.

Borders Tab
- **Border properties** part is:

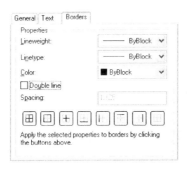

- Control the following settings:
  - Sepcify the **Lineweight**, **Linetype**, and **Color** of the borders (ByBlock or specify from the list the desired value).
  - Specify whether you want the border single line (default) or double line. If you specify double line, specify the spacing.
  - Set the type of border (from inside, outside, etc.).
- You can select from many styles available one to make Current, so next time you use the **Table** command you will use this style.
- You can also select one of the existing styles and make any type of modification. You will get the same dialog box you got when you created this style.

CREATING A TABLE STYLE (METRIC)

Workshop 15-A

1. Start AutoCAD 2008, and open the file: **Small_Villa_Ground_ Floor_Plan_Metric_Workshop_15.dwg**. (If you solved the previous workshop correctly, open your file.)

2. Create a new Table Style based on Standard and call it **Door Schedule**,
   a. For Data, Text Style = **Inside_Annot** and Alignment = **Middle Center**
   b. For Column Heads, Text Style = **Inside_Annot** and Alignment = **Middle Center**
   c. For Title, Text Height = **500**
   d. Cell margins, Horizontal = **100**, Vertical = **100**
3. Save the file and close it.

CREATING A TABLE STYLE (IMPERIAL)

Workshop 15-B

1. Start AutoCAD 2008, and open the file: **Small_Villa_Ground_Floor_Plan_Imperial_Workshop_15.dwg**. (If you solved the previous workshop correctly, open your file.)
2. Create a new Table Style based on Standard and call it **Door Schedule**,
   a. For Data, Text Style = **Inside_Annot** and Alignment = **Middle Center**
   b. For Column Heads, Text Style = **Inside_Annot** and Alignment = **Middle Center**
   c. For Title, Text Height = **1′–8″**
   d. Cell margins, Horizontal = **4″**, Vertical = **4″**
3. Save the file and close it.

TABLE COMMAND

- To insert a table in an AutoCAD drawing using a predefined style.
- The user will specify the number of columns and rows, and will fill the cells with the desired data.
- To issue this command use one of the following methods:

  - From the dashboard and using the **Tables** panel, click the **Table** button.
  - From menus select **Draw/Table**.
  - Type **table** in the Command window.

- The following dialog box will appear:

FIGURE 8-12

- Select the predefined **Table style name**.
- If you didn't create a Table style before this step, simply click the small button beside the list so you can start creating the desired Table style.
- Specify the **Insert options**. You have three choices:
  - Start from empty table—normally this option should be used.
  - From a data link—to bring in data from spreadsheets like MS Excel.
  - From object data in the drawing (Data Extraction) —only if you have block attributes.
- There are two insertion methods:
  - Sepcify insertion point
  - Specify window

Specify Insertion Point

- If you use this method, you specify the upper left corner of the table, and accordingly set up the following data:
  - Number of columns
  - Column width
  - Number of rows (without Title and Column Heads)
  - Row height (in lines)
- Click **OK** and AutoCAD will prompt:

```
Specify insertion point:
```

- Specify the upper left corner of the table and the table will appear ready to fill the data in each row. You will fill first the title, the column headers,

Specify Window
then the data. You can move between rows using the [Tab] key on the keyboard to go to the next cell and [Shift]+[Tab] to go back to the previous cell.

- If you use this method, you will be asked later to specify a window, hence you specify a total height and a total width. Accordingly, fill the following data:
  - Either specify the number of columns and the column width will be calculated automatically (*total width/number of columns*), or specify the column width, and the number of columns will be calculated automatically (*total width/single column width*).
  - The same thing goes for rows. Either specify the number of rows and the row height will be calculated automatically (*total height/number of rows*), or specify the row height and the number of rows will be calculated automatically (*total height/single row height*).
- Click **OK** and AutoCAD will prompt:

```
Specify first corner:
Specify second corner:
```

- Specify two opposite corners, then the table will be available for your input just like we did in the previous method.

- To edit cell content, simply double-click the cell, and it will be available for editing.

INSERTING TABLES (METRIC)

Workshop 16-A

1. Start AutoCAD 2008, and open the file: **Small_Villa_Ground_Floor_Plan_Metric_Workshop_16.dwg**. (If you solved the previous workshop correctly, open your file.)
2. Make the **Text** layer the current layer.
3. Looking at the picture below and using the Table style **Door Schedule**, add a table just like the one below using the following:
   a. Specify insertion point
   b. Columns = **5**
   c. Column Width = **2000**
   d. Data Rows = **4**
   e. Row Height = **1** Line(s)

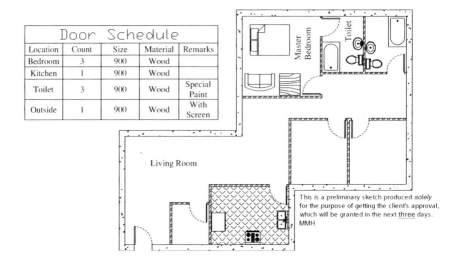

Ground Floor Plan

4. Save the file and close it.

INSERTING TABLES (IMPERIAL)

 Workshop 16-B

1. Start AutoCAD 2008, and open the file: **Small_Villa_Ground_Floor_Plan_Imperial_Workshop_16.dwg**. (If you solved the previous workshop correctly, open your file.)
2. Make the **Text** layer the current layer.
3. Looking at the picture below and using the Table style **Door Schedule**, add a table just like the one below using the following:
    a. Specify insertion point
    b. Columns = **5**
    c. Column Width = **6′–8″**
    d. Data Rows = **4**
    e. Row Height = **1** Line(s)

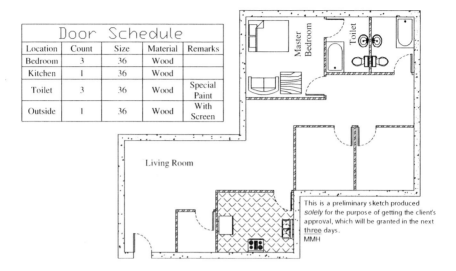

4. Save the file and close it.

CHAPTER REVIEW

1. The height mentioned in the Text Style is for lowercase letters.
   a. True
   b. False
2. There are two types of fonts in AutoCAD _____ and _____.
3. There is no relation between Text style and Table style.
   a. True
   b. False
4. While you are in the In-place text editor you can't:
   a. Import any .txt file
   b. Format text
   c. Change the indents
   d. Bring on an MS Word document as OLE
5. One of the following is not an editing method for text:
   a. Double-clicking.
   b. Right-click and select the Mtext Edit option.

c. The command ddedit for both types and the command mtedit for multiline text.
　　d. The command mtextedit.

CHAPTER REVIEW ANSWERS

1. b
2. .shx, .ttf
3. b
4. d
5. d

Chapter **9**

DIMENSIONING YOUR DRAWING

In This Chapter

- Dimension types
- How to create Parent and Child dimension styles
- How to control dimension styles
- The different types of dimensioning commands
- Editing dimensions and editing dimensions with Grips

INTRODUCTION

- Dimensioning in AutoCAD is a semiautomatic process, as users contribute to part of the job and AutoCAD does the rest.
- Users (in Linear dimensioning for instance) specify three points, the first and the second is the length to be dimensioned and the third is the position of the dimension line.
- Accordingly, AutoCAD will automatically generate the Dimension block, just like the illustration below:

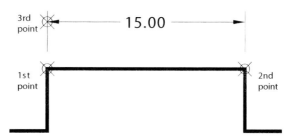

- Dimension block consists of four parts. They are:
  - Dimension line
  - Extension lines
  - Arrowheads
  - Dimension Text

- See the following illustration:

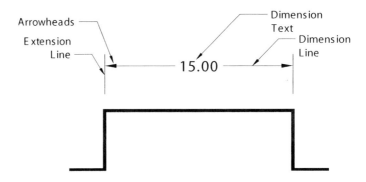

- The dimensioning process goes into two phases:
  - Create Dimension style(s).
  - Putting dimensions over your drawing.
- Dimension style will control the appearance of the Dimension Block, so each can set up the style as they wish and according to their standards.
- Usually creating a dimension style is a lengthy and tedious job, but it is only done once, which allows the user to focus on the other job—putting the dimension over the drawing.

DIMENSION TYPES

- These are the dimension types AutoCAD can support:

Linear and Aligned

- Check the following illustration:

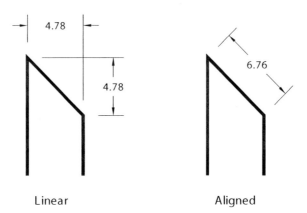

Arc Length, Radius, and Diameter ■ Check the illustration below:

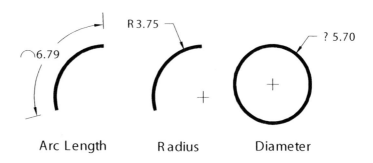

Angular ■ Check the illustration below:

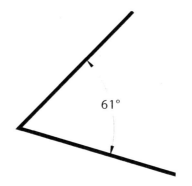

Continuous ■ Check the illustration below:

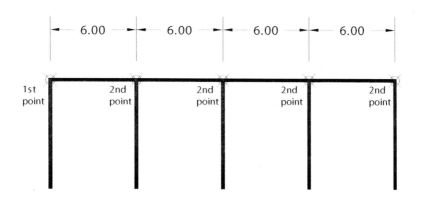

Baseline — Check the illustration below:

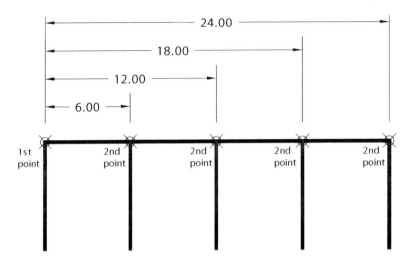

Ordinate — Check the illustration below:

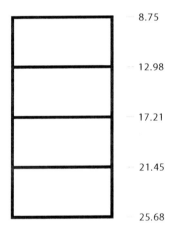

Quick Leader — Check the illustration below:

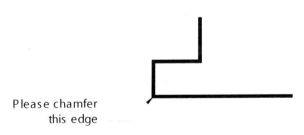

DIMENSION STYLE: THE FIRST STEP

- To issue this command, use one of the following methods:
  - From the dashboard and using the **Dimensions** panel, or from **Dimension** toolbar, click the **Dimension Style** button.
  - From menus select **Format/Dimension Style**.
  - Type **dimstyle** in the Command window.
- The following dialog box will appear:

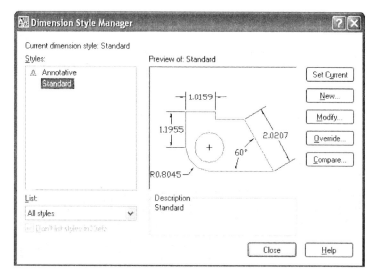

FIGURE 9-1

- By default there will be a premade dimension style called **Standard**.
- Users can modify this style or create their own (preferable).

- To create a new style, click, New button. The following dialog box will appear:

FIGURE 9-2

- Type in the name of your new style following the same naming convention of layers.
- Select the **Start With** style (you will start with a copy of this style).
- Keep **Annotative** off for now.
- By default, the changes you make will be implemented for all types of dimesions, but you can create a new dimension style that will affect certain types of dimensions.

- Click the **Continue** button, then you can start modifying the settings.
- We will cover each tab of the Dimension style dialog box in the following pages.

- In this dialog box, whenever you find the **Color** setting, leave it as is. We should control our colors through layers and not through individual objects. This also applies for **Linetype** and **Lineweight**.

DIMENSION STYLE: LINES TAB

- The first tab in the Dimension style dialog box is Lines, where we will control Dimension lines and Extension lines. It looks like the following:

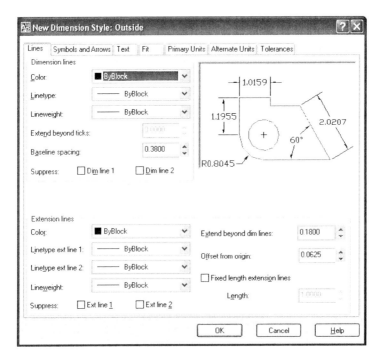

FIGURE 9-3

- Under **Dimension lines** you can control the following settings:
  - The **Color**, **Linetype**, and **Lineweight** of the Dimension line.
  - **Extended beyond ticks** (in order to edit this value, go to the **Symbols and Arrows** tab and set the **Arrowhead** to **Architectural tick** or **Oblique**). See the following illustration:

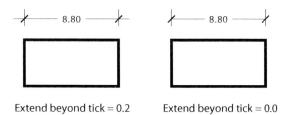

Extend beyond tick = 0.2   Extend beyond tick = 0.0

  - **Baseline spacing**, see the illustration below:

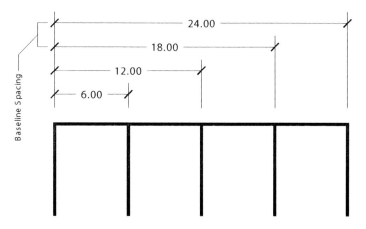

  - Select to **Suppress Dim line 1**, **Dim line 2**, for one of them or for both. See the illustration below:

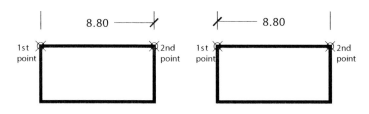

Suppress Dim line 1=True   Suppress Dim line 2=True

- Under **Extension lines** you can control the following settings:
  - The **Color**, **Linetype**, and **Lineweight** of the Extension lines.

- Select to **Suppress Ext line 1**, **Ext line 2**, for one of them or for both. See the illustration below:

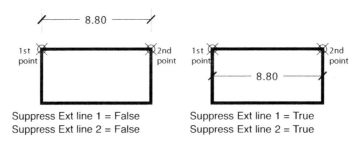

- Specify **Extend beyond dim lines** and **Offset from origin**. Check the illustration below:

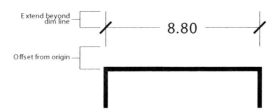

- Select to set a **Fixed length extension line**. You have to specify the **Length**. In order to understand this issue, see the below example:

Example
- This is an example to clarify **Fixed length extension lines**:
  - Assume we have the following case and we want to put a linear dimension

- Before AutoCAD 2006, the only choice was as follows:

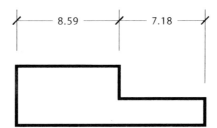

- As of AutoCAD 2008, a new choice called **Fixed length extension lines** is available to get the following:

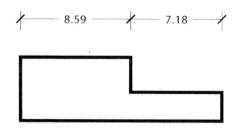

 - The Length you specify will be calculated from the dimension origin up to the dimension line.

DIMENSION STYLE: SYMBOLS AND ARROWS TAB

- Click the **Symbols and Arrows** tab and you will see the following:

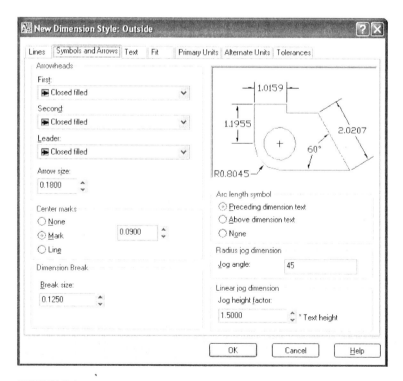

FIGURE 9-4

- Under **Arrowheads** you can control the following settings:
  - The shape of the **First** arrowhead.
  - The shape of the **Second** arrowhead.
  - The shape of the arrowhead to be used in the **Leader**.
  - The **Size** of the arrowhead.

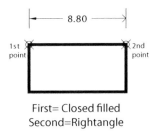

- If you change the first, the second will change automatically, but if you change the second the first will not change.
- Under **Center marks** you can control the following settings:
  - Select whether to show the Center mark or not, or to show centerlines.
  - Set the **Size** of the Center mark.

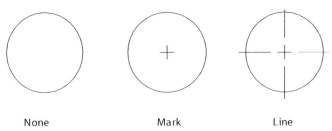

- Under **Arc length symbol** you can control whether to show the symbol **Preceding** the dimesion text, **Above** the dimesion text, or **None**. Check the illustration below:

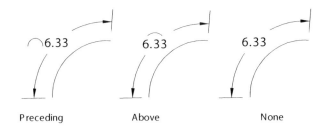

- Under **Dimension Break**, set the **Break size**, which is the width of the break of the dimension lines in a dimension break.

- Under **Radius dimension jog**, set the value of the **Jog angle**. See the illustration below:

- Under **Linear Jog dimension**, set the **Jog height factor** as a percentage of the of the text height. See the illustration below.

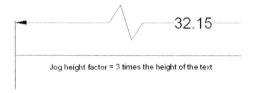

DIMENSION STYLE: TEXT TAB

- Click the **Text** tab and you will see the following:

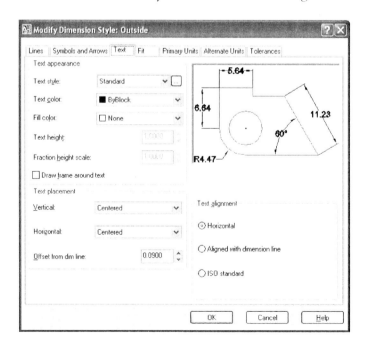

FIGURE 9-5

- Under **Text appearance** you can control the following settings:
  - Select the desired premade **Text style** to be used to write the dimension text. If you didn't create a Text style prior to this step, you can click the three dots button and create it right now.
  - Specify the **Text color**.
  - Specify the **Fill color**, which is the background color for the dimension text.
  - Input the **Text height** (this is only applicable if the assigned Text style has a text height = 0.0).
  - If you go to the **Primary Units** tab and assign the **Unit format** to be **Architectural** or **Fractional**, then the dimension text will appear something like 1 1/4. The question here is whether you want the fraction to appear with less height than the ordinary number, set the **Fraction height scale** accordingly.
  - Select whether to **Draw frame around** the dimension **text** or not.
- As an introduction to **Text placement**, check the following illustration:

Vertical Placement = Centered

Horizontal Placement = Centered

- Under **Text placement** you can control the following settings:
  - Select the **Vertical** placement. You have 4 choices: Centered, Above, Outside, and JIS (Japan Industrial Standard). Check the illustration below:

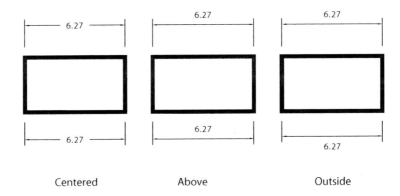

- As for **Above**, set the **Offset from dim line**, which is the distance between the dimesion line, and the baseline of the dimension text. Check the illustration below:

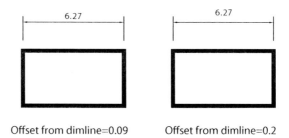

- Select the **Horizontal** placement. You have 5 choices: Centered, At Ext Line 1, At Ext Line 2, Over Ext Line 1, and Over Ext Line 2. Check the illustration below:

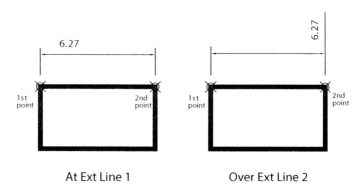

- Under **Text alignment**, you can control whether the text will be **Horizontal** always, **Alighned with dimension line**, or according to **ISO standard**. Check the following illustration:

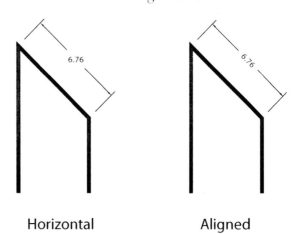

■ The only difference between **Aligned with dimesion line** and **ISO standard** is with the Radius and Diameter types, where the first is considered like the other types aligned with the dimension line and ISO standard considers it as a special case to be horizontal.

DIMENSION STYLE: FIT TAB

■ Click the **Fit** tab and you will see the following:

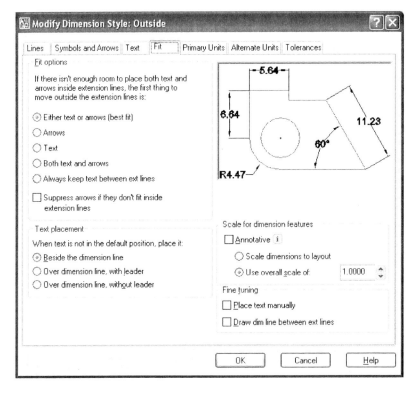

FIGURE 9-6

■ As an introduction to **Fit** tab, users will notice that there are three things within the two Extension lines, which are: Dimension line, Arrowheads, and Dimension text. AutoCAD will put them all inside the two Extension lines when the distance is comfortable enough for all of them. But if there is not enough room? That is discussed next.

- Under **Fit options**, you can control the following settings:
  - Select one of the five options to decide how AutoCAD will treat Arrowheads and Dimension text.
  - Select whether to **Suppress arrows if they don't fit inside extension lines** or not.
- Under **Text placement**, you can control the placement of the text. If it doesn't fit inside the extension lines you have three options to choose from: **Beside the dimension line, Over dimension line with leader,** and **Over dimension line without leader**. Check the illustration below:

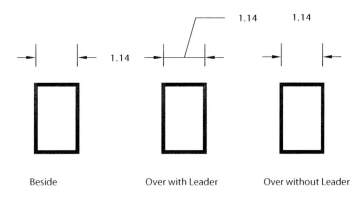

Beside   Over with Leader   Over without Leader

- Under **Scale for dimension features**, you can control the following settings:
  - Keep **Annotative** off for now.
  - Set the **Scale dimension to layout** (we will discuss layouts in the next chapter)
  - For any distance, length, or size, users will input a value. **Use overall scale of** is a setting that will magnify or shrink the whole values in one shot. This will not affect the distance measured.
- Under **Fine tunning**, you can control the following settings:
  - If you don't trust AutoCAD to place your text in the right place you can choose to let AutoCAD allow you to **Place** the dimension **text manually**.

- Also, you can choose to force AutoCAD to **Draw the dimension line between extension lines**, whether the distance is appropriate or not.

DIMENSION STYLE: PRIMARY UNITS TAB

- Click the **Primary Units** tab and you will see the following:

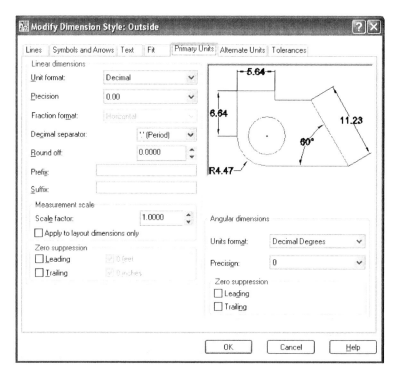

FIGURE 9-7

- As an introduction to the **Primary units**, let us assume that your client wants the dimensions in Decimal and one other subcontractor wants them in Architectural. The solution for that is to show two numbers for each dimension, the first will be the **Primary Units** and the

second will be the **Alternate Units**. In this tab we will cover the **Primary Units**.
- Under **Linear dimensions**, you can control the following settings:
  - Choose the **Unit format**, select one of six formats.
  - Select the **Precision** of the unit format selected.
  - If you select as the Unit format Architectural or Fractional, then specify the **Fraction format**, which are: **Horizontal**, **Diagonal**, and **Not Stacked**.
  - If you select **Decimal**, then specify the **Decimal Separator**, which are: **Period**, **Comma**, and **Space**.
  - Specify the **Round off** number. If you select, for instance, 0.5, then AutoCAD will round off any dimension to the nearest 0.5.
  - Input the **Prefix** and/or the **Suffix**. To illustrate this part, see below:

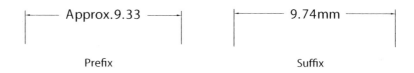

- Under **Measurement scale**, you can control the following:
  - Input the **Scale factor**. To explain the importance of this setting, let's take the following example: Assume we have a drawing that uses the unit millimeter, hence a length of 10 m, will be 10,000. But in the dimension, we want the value 10 to appear and not 10,000, hence we set the Scale factor to be 0.001.
  - Select to **Apply to layout dimesions only** (we will discuss the layouts in the next chapter).
- Under **Zero suppression**, select to suppress the **Leading** and/or the **Trailing** zeros. See the illustraton below:

- Under **Angular dimensions**, select the **Unit format** and the **Precision**.
- Under **Zero suppression**, select to suppress the **Leading** and/or the **Trailing** zeros for the angular measurements.

DIMENSION STYLE: ALTERNATE UNITS TAB

- Click the **Alternate Units** tab and you will see the following:

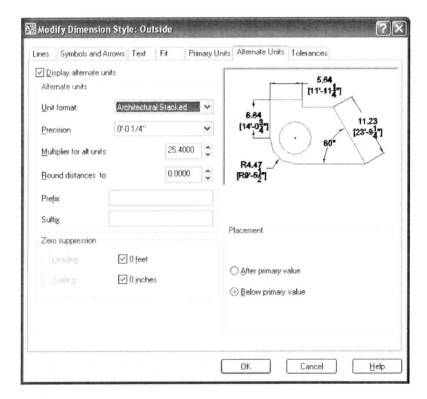

FIGURE 9-8

- If you want two numbers to appear for each dimension, click on **Display alternate units**.
- Specify the Alternate **Unit format**, its **Precision**, the **Multiplier for all units** value, the **Round distance**, the **Prefix**, the **Siffix**, and the **Zero suppression** criteria.
- Specify whether to show the alternate units **After primary value** or **Below primary value**.

DIMENSION STYLE: TOLERANCES TAB

- Click the **Tolerances** tab and you will see the following:

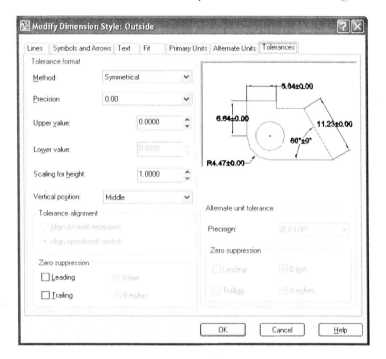

FIGURE 9-9

- There are several ways to show the tolerances. They are:
  - None
  - Symmetrical
  - Deviation
  - Limits
  - Basic
- Check the illustrations on the next page.

None
- Check the illustration below:

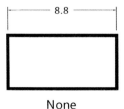

None

Symmetrical ■ Check the illustration below:

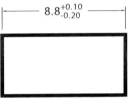

Symmetrical

Deviation ■ Check the illustration below:

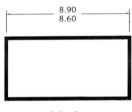

Deviation

Limits ■ Check the illustration below:

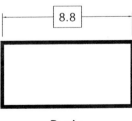

Limits

Basic ■ Check the illustration below:

8.8

Basic

- Under **Tolerance format**, you can control the following:
  - Specify the desired **Method**, one of the above mentioned methods.
  - Specify the **Precision** of the numbers to be shown.
  - If you select **Symmetrical**, specify the **Upper value**. For **Deviation** and **Limits** specify the **Upper** and **Lower value**.
  - If you select the Tolerance values to appear smaller than the dimension text, specify **Scaling for height**.
  - Specify the **Vertical position** of the dimension text with regard to the tolerance values whether **Bottom**, **Middle**, or **Top**.
- If you are showing **Alternate units**, specify the **Precision** of the numbers under the **Alternate units tolerance**.
- Accordingly, specify the **Zero suppression** for both the **Primary Units** tolerance and the **Alternate units** tolerance.

DIMENSION STYLE: CREATING A CHILD STYLE

- One of the cases users may face is: You need a dimension style identical to almost all types of dimensions except for Diameter, which you want a little bit different.
- In the above case, we create a dimension style for all types (we call Parent), then we create a child dimension style from it.
- Follow the following procedure:
  - Create your parent dimension style.
  - Select it from the list in the dimension style dialog box.
  - Click the **New** button to create a new style.
  - The following dialog box will appear:

FIGURE 9-10

- For **Use for** select **Radius** (as an example). The dialog box will change to:

FIGURE 9-11

- Now click **Continue** and make the changes you want. These changes will affect only the Radius dimensions.
- In the **Dimension Style** dialog box you will see something like:

```
out-side
    Radial
Standard
```

CONTROLLING DIMENSION STYLES

- Once you create more than one dimension style you can control these dimension styles using the following buttons:

- To set the selected dimension style as the current style.
- Other ways of setting a dimension style as current are:
  - From the **Styles** toolbar, select the desired dimension style:

  - From the **Dimension** toolbar, select the desired dimension style:

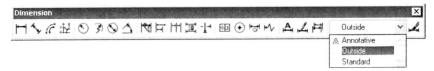

- From the dashboard and using the **Dimensions** panel, set the current dimension styles using the pop-up list:

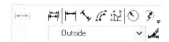

- In the Dimension style dialog box, double-click the name of the desired dimension style and it will become the current dimension style.

- To modify the selected dimension style. The dialog box that we discussed in creating the dimension style will appear again for the purpose of changing any desired setting.

To Delete

- In order to delete a dimension style there are two conditions:
  - It should not be used in the current drawing.
  - It should not be the current dimension style.
- If these two conditions are fulfilled, then select the desired dimension style to delete and press [Del] on the keyboard. The following dialog box will appear:

FIGURE 9-12

- If you click **Yes**, it will be deleted, and if you click **No**, the whole process will be canceled.

CREATING DIMENSION STYLES (METRIC)

Workshop 17-A

1. Start AutoCAD 2008, and open the file: **Small_Villa_Ground_Floor_Plan_Metric_Workshop_17.dwg**. (If you solved the previous workshop correctly, open your file.)

2. Create a new dimension style and name it **Outside**, starting from **Standard**, and Use for **All dimensions**. (Anything not mentioned below leave on the default value or setting.)
3. Under **Line** make the following changes:
   a. Extend beyond dime line = 0.25
   b. Offset from origin = 0.15
4. Under **Symbols and Arrows** make the following changes:
   a. Arrowhead, First = Oblique
   b. Arrow size = 0.25
5. Under **Text** make the following changes:
   a. Text style = Dimension
   b. Text placement, Vertical = Above
   c. Text alignment = Aligned with dimension line
6. Under **Fit** make the following changes:
   a. Use overall scale = 1000
7. Under **Primary Units** make the following changes:
   a. Linear dimension, Precision = 0.00
   b. Suffix = m
   c. Scale Factor = 0.001
8. Create a new style and name it **Inside**, starting from **Outside**, and use for **All dimensions**.
9. Under **Lines** make the following changes:
   a. Extension lines, Suppress Ext line 1 = on, Ext line 2 = on
10. Under **Symbols and Arrows** make the following changes:
    a. Arrow size = 0.20
11. Under **Text** make the following changes:
    a. Text style = Standard
    b. Text height = 0.25
12. Under **Fit** make the following changes:
    a. Fine tuning, Place text manually = on
13. Make a child dimension style from **Outside** for **Radius dimensions**.
14. Under **Symbols and Arrows** make the following changes:
    a. Arrowheads, Second = Closed filled
15. Under **Text** make the following changes:
    a. Text alignment = ISO Standard.
16. Save the file and close it.

CREATING DIMENSION STYLES (IMPERIAL)

 Workshop 17-B

1. Start AutoCAD 2008, and open the file: **Small_Villa_Ground_Floor_ Plan_Imperial_Workshop_17.dwg**. (If you solved the previous workshop correctly, open your file.)
2. Create a new dimension style and name it **Outside**, starting from **Standard**, and Use for **All dimensions**. (Anything not mentioned below leave on the default value or setting.)
3. Under **Line** make the following changes:
    a. Extend beyond dime line = 3/4"
    b. Offset from origin = 1/2"
4. Under **Symbols and Arrows** make the following changes:
    a. Arrowhead, First = Oblique
    b. Arrow size = 3/4"
5. Under **Text** make the following changes:
    a. Text style = Dimension
    b. Text placement, Vertical = Above
    c. Text alignment = Aligned with dimension line
6. Under **Fit** make the following changes:
    a. Use overall scale = 12
7. Create a new style and name it **Inside**, starting from **Outside**, and use for **All dimensions**.
8. Under **Lines** make the following changes:
    a. Extension lines, Suppress Ext line 1 = on, Ext line 2 = on
9. Under **Text** make the following changes:
    a. Text style = Standard (change the font to be Arial)
    b. Text height = 3/4"
10. Under **Fit** make the following changes:
    a. Fine tuning, Place text manually = on
11. Make a child dimension style from **Outside** for **Radius dimensions**.
12. Under **Symbols and Arrows** make the following changes:
    a. Arrowheads, Second = Closed filled
13. Under **Text** make the following changes:
    a. Text alignment = ISO Standard.
14. Save the file and close it.

DIMENSIONING COMMANDS: LINEAR

- To create a horizontal or vertical dimension.
- To issue this command, use one of the following methods:
  - From the dashboard and using the **Dimensions** panel, click the **Linear** button. Or, from the **Dimension** toolbar, select the **Linear** button.
  - From menus select **Dimension/Linear**.
  - Type **dimlinear** in the Command window.
- The following prompt will appear:

  Specify first extension line origin or <select object>: *(Specify the first point)*
  Specify second extension line origin: *(Specify the second point)*
  Specify dimension line location or [Mtext/Text/Angle/Horizontal/Vertical/Rotated]: *(Specify the location of the dimension line)*

- Users have to undergo three steps, which are:
  - Specify the first point of the dimension distance to be measured.
  - Specify the second point of the dimension distance to be measured.
  - Specify the location of the dimension block by specifying the location of the dimension line.
- The following is the result:

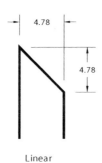

Linear

- You can use the other options available, which are:
  - Mtext
  - Text
  - Angle
  - Horizontal
  - Vertical
  - Rotated

Mtext — To edit the measured distance in MTEXT command.
Text — To edit the measured distance in DTEXT command.

Angle ■ To set up the angle of text.
Horizontal ■ To create a horizontal dimension and not vertical.
Vertical ■ To create a vertical dimension and not horizontal.
Rotated ■ To create a dimension line parallel to another angle given by the user. As in the case of projecting a distance over another angle.

DIMENSIONING COMMANDS: ALIGNED

- To create a dimension parallel to the measured distance.
- To issue this command, use one of the following methods:
  - From the dashboard and using the **Dimensions** panel, click the **Aligned** button. Or, from the **Dimension** toolbar, select the **Aligned** button.
  - From menus select **Dimension/Aligned**.
  - Type **dimaligned** in the Command window.
- The following prompt will appear:

  ```
  Specify first extension line origin or <select
  object>: (Specify the first point)
  Specify second extension line origin: (Specify
  the second point)
  Specify dimension line location or
  [Mtext/Text/Angle]: (Specify the location of the
  dimension line)
  ```

- Users have to undergo three steps, which are:
  - Specify the first point of the dimension distance to be measured.
  - Specify the second point of the dimension distance to be measured.
  - Specify the location of the dimension block by specifying the location of the dimension line.
- The following is the result:

Aligned

- The rest of the options are just like **Linear** command.

LINEAR AND ALIGNED DIMENSIONS

Exercise 29

1. Start AutoCAD 2008.
2. Open the file **Exercise_29.dwg**.
3. Create a new layer and call it **Dimension**, Color = **White**, and make it current.
4. Make the following modification to the current dimension style (i.e., Standard):
   a. Under **Symbols and Arrows**, Arrow size = **0.10**
   b. Under **Text**, Text height = **0.15**
   c. Under **Primary Units**, make the Linear Precision = **0.00**
5. Make the linear and aligned dimensions as shown:

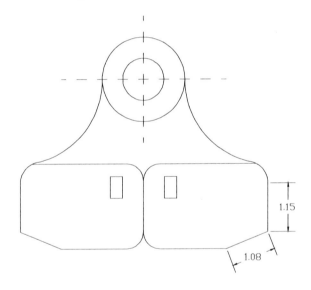

6. Save the file and close it.

DIMENSIONING COMMANDS: ARC LENGTH

- To create a dimension showing the length of a selected arc.
- To issue this command, use one of the following methods:
  - From the dashboard and using the **Dimensions** panel, click the **Arc Length** button. Or, from the **Dimension** toolbar, select the **Arc Length** button.

- From menus, select **Dimension/Arc Length**.
- Type **dimarc** in the Command window.
- The following prompt will appear:

```
Select arc or polyline arc segment: (Select the
desired arc)
Specify arc length dimension location, or
[Mtext/Text/Angle/Partial/Leader]: (Specify the
location of the dimension block)
```

- Users have to undergo two steps, which are:
  - Select the desired arc.
  - Specify the location of the dimension block.
- The following is the result:

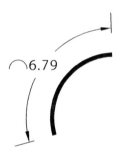

- The options Mtext, Text, and Angle, were already discussed in the Linear command.

Partial
- If you want Arc Length to measure part of the arc and not all of the arc, specify two points on the arc accordingly, the result looks something like:

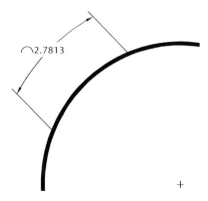

Leader ■ For an arc with an angle more than 180 you may add a leader, just like the example below:

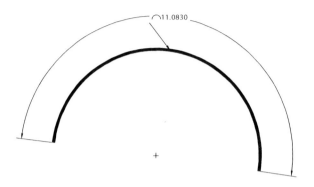

DIMENSIONING COMMANDS: ORDINATE

- To add several measurements to objects relative to a certain point.
- To issue this command, use one of the following methods:
  - From the dashboard and using the **Dimensions** panel, click the **Ordinate** button. Or, from the **Dimension** toolbar, select the **Ordinate** button.
  - From menus, select **Dimension/Ordinate**.
  - Type **dimordinate** in the Command window.
- Ordinate command allows the user to set dimensions relative to a datum, either in X or in Y. Check the illustration below:

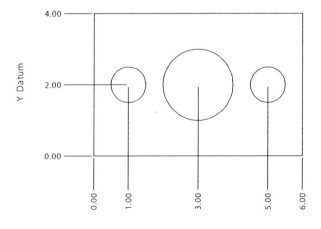

- You have to change the UCS origin location to the desired location so the readings will be right, otherwise, the values will be relative to the current 0,0.
- The prompts of the command will be as follows:

   Specify feature location: *(Click on the desired point)*

   Specify leader endpoint or [Xdatum/Ydatum/Mtext/Text/Angle]: *(Specify the dimension location)*

- By default, when you select a point, you may go in the direction of X or in Y. If you want the Ordinate command to go specifically in the direction of X, then select the **Xdatum** option, and select **Ydatum** if you want to go specifically in the direction of Y.
- The rest of the options were discussed in the **Linear** command.

DIMENSIONING COMMANDS: RADIUS

- To put Radius dimension on an arc and/or circle.
- To issue this command, use one of the following methods:
  - From the dashboard and using the **Dimensions** panel, click the **Radius** button. Or, from the **Dimension** toolbar, select the **Radius** button.
  - From menus select **Dimension/Radius**
  - Type **dimradius** in the Command window
- The following prompts will appear:

   Select arc or circle: *(Select the desired arc or circle)*

   Specify dimension line location or [Mtext/Text/Angle]: *(Specify the location of the dimension block)*

- Users have to undergo two steps, which are:
  - Select the desired arc or circle.
  - Specify the location of the dimension block.
- This is the result you will get:

DIMENSIONING COMMANDS: JOGGED

- When the arc is big and its center may be outside the current view, users may need to use the **Jogged** radius instead of the normal radius.
- To issue this command, use one of the following methods:
  - From the dashboard and using the **Dimensions** panel, click the **Jogged** button. Or, from the **Dimension** toolbar, select the **Jogged** button.
  - From menus select **Dimension/Jogged**.
  - Type **dimjogged** in the Command window.
- The following prompts will appear:

  Select arc or circle: *(Select the desired arc or circle)*
  Specify center location override: *(Specify the imaginary new location of the center)*
  Specify dimension line location or [Mtext/Text /Angle]: *(Specify the dimension block location)*
  Specify jog location: *(Specify the location of the jog)*

- This is the result you will get:

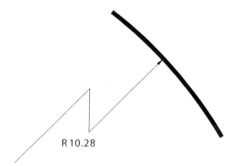

DIMENSIONING COMMANDS: DIAMETER

- To put diameter dimension on an arc and/or circle.
- To issue this command, use one of the following methods:
  - From the dashboard and using the **Dimensions** panel, click the **Diameter** button. Or, from the **Dimension** toolbar, select the **Diameter** button.
  - From menus select **Dimension/Diameter**.
  - Type **dimdiameter** in the Command window.

- The following prompts will appear:

  Select arc or circle: *(Select the desired arc or circle)*
  Specify dimension line location or [Mtext/Text/Angle]: *(Specify the dimension block location)*

- This is the result you will get:

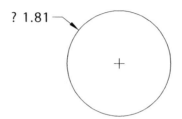

DIMENSIONING COMMANDS: ANGULAR

- To issue this command, use one of the following methods:
  - From the dashboard and using the **Dimensions** panel, click the **Angular** button. Or, from the **Dimension** toolbar, select the **Angular** button.
  - From menus select **Dimension/Angular**.
  - Type **dimangular** in the Command window.
- There are four ways to place an angular dimension in AutoCAD. They are:
  - Select an arc, AutoCAD will measure the included angle or the outside angle.
  - Select a circle, the position that you select the circle will be the first point, the center of the circle will be the second point. AutoCAD will ask the user to specify any point on the diameter of the circle, and will place the angle accordingly.
  - Select two lines, AutoCAD will measure the inside angle or the outside angle.
  - Select a vertex, which will be considered as the center point, then AutoCAD will ask you to specify two points and measure the inside angle or the outside angle.
- The following prompts will appear:

  Select arc, circle, line, or <specify vertex>: *(Select the desired method as discussed above—assume we select arc)*

```
Specify dimension arc line location or [Mtext/
Text/Angle]: (Specify the dimension block location)
```

- This is the result you will get:

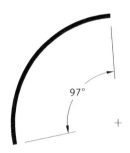

ARC LENGTH, RADIUS, JOGGED, DIAMETER, AND ANGULAR DIMENSIONS

 Exercise 30

1. Start AutoCAD 2008.
2. Open the file **Exercise_30.dwg**.
3. Make the following modification to the current dimension style (i.e., Standard):
   a. Under **Text**, change the current text style to use **Arial** font.
4. Do the four types of dimensions as shown:

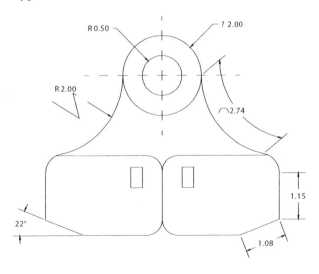

5. Save the file and close it.

DIMENSIONING COMMANDS: CONTINUE

- After you put a dimension in your drawing (i.e., Linear, Aligned, Angular, etc.), you can ask AutoCAD to continue using the same type and to allocate it along the first one.
- Using this command will allow the user to put lots of dimensions swiftly.
- To issue this command, use one of the following methods:
  - From the dashboard and using the **Dimensions** panel, click the **Continue** button. Or, from the **Dimension** toolbar, select the **Continue** button.
  - From menus select **Dimension/Continue**.
  - Type **dimcontinue** in the Command window.

If No Dimension was Created in this Session
- The following prompts will appear:

```
Select continued dimension: (Select either Linear,
Aligned, Ordinate, or Angular)
```

- AutoCAD will consider the selected dimension as the base dimension and will continue accordingly.

If a Dimension was Created in this Session
- The following prompt will appear:

```
Specify a second extension line origin or [Undo/
Select] <Select>: (Specify the second point of
the last Linear, Aligned, Ordinate, or Angular,
or select an existing dimension)
```

- AutoCAD will give you the ability to do one of three things:
  - If you already put a linear dimension (for example), then you can continue by specifying the second point, considering the second point of the first dimension is the first point of the continuing dimension.
  - Also, you can select an existing dimension and continue from there.
  - Or, you can undo the last continue dimension.
- Check the following illustration:

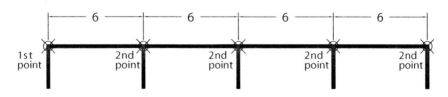

DIMENSIONING COMMANDS: BASELINE

- Just like Continue except the dimensions will always be related to the first point the user selected.
  - From the dashboard and using the **Dimensions** panel, click the **Baseline** button. Or, from the **Dimension** toolbar, select the **Baseline** button.
  - From menus select **Dimension/Baseline**.
  - Type **dimbaseline** in the Command window.
- All the prompts and the procedures are identical to the **Continue** command.
- Check the following illustration:

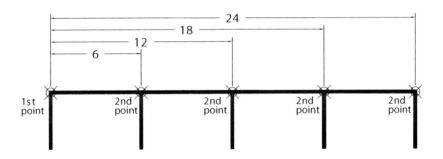

CONTINUOUS AND BASELINE DIMENSIONS

Exercise 31

1. Start AutoCAD 2008.
2. Open the file **Exercise_31.dwg**.
3. Create a new layer, call it **Dimension_2**, and make it current, freeze layer **Dimension**.
4. Do the Continuous and Baseline as shown:

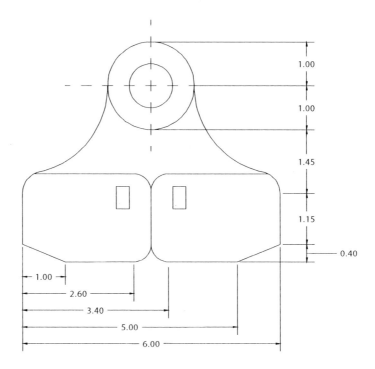

5. Save the file and close it.

DIMENSIONING COMMANDS: QUICK DIMENSION

- To place a group of dimensions in a single step.
  - From the dashboard and using the **Dimensions** panel, click the **Quick Dimension** button. Or, from the **Dimension** toolbar, select the **Quick Dimension** button.
  - From menus select **Dimension/Quick Dimension**.
  - Type **qdim** in the Command window.
- The following prompt will appear:

```
Select geometry to dimension: (Either by click-
ing, Window, or Crossing)
Specify dimension line position, or
[Continuous/Staggered/Baseline/Ordinate/Radius/
Diameter/datumPoint/Edit/settings] <Continuous>:
```

- At this prompt you can right-click and the following shortcut menu will appear:

- From this shortcut menu you can select the proper dimension type. They are: Continuous, Staggered, Baseline, Ordinate, Radius, or Diameter.
- Select the type, then specify the dimension line location. A group of dimensions will be placed in a single step.
- Check the **Staggered** example below:

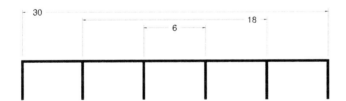

DIMENSIONING COMMANDS: QUICK LEADER

- To create a leader pointing to part of your drawing with text.
- In order to issue the command, use one of the following methods:
  - From the dashboard and using the **Dimensions** panel, click the **Quick Leader** button. Or, from the **Dimension** toolbar, select the **Quick Leader** button.
  - From menus select **Dimension/Leader**.
  - Type **qleader** in the Command window.

- The following prompt will appear:

  Specify first leader point, or [Settings] <Settings>: *(Specify the first point which the arrow will point to)*
  Specify next point: *(Specify the second point)*
  Specify next point: *(Specify the third point)*
  Specify text width <0>: *(Specify the width of the text if you want to constraint the text or else press[Enter])*
  Enter first line of annotation text <Mtext>: *(Start inputting the text as you wish)*

- You will get something similar to the following:

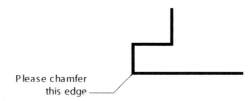

- After inputting a leader you can click the text. The four grips will appear in order to edit the area the text occupies, just like we did in the Text lesson.
- When you input the leader, if you click the three points starting from left to right, the text will be Left justified.
- When you input the leader, if you click the three points starting from right to left, the text will be Right justified.

DIMENSIONING COMMANDS: CENTER MARK

- To place a Center Mark for arcs and circles.
- To issue this command, use one of the following methods:
  - From the dashboard and using the **Dimensions** panel, click the **Center Mark** button. Or, from the **Dimension** toolbar, select the **Center Mark** button.
  - From menus, select **Dimension/Center Mark**.
  - Type **dimcenter** in the Command window.
- The following prompt will appear:

  Select arc or circle: *(Select the desired arc, or circle)*

- You will get something similar to the following:

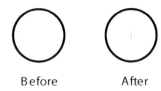

Before   After

DIMENSION BLOCKS AND GRIPS

- You can edit dimension blocks by Grips.
- If you click a dimension block, 5 grips will appear, just like the following:

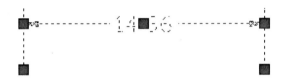

- From the above illustration, we can see that grips appear in the following places:
  - The two ends of the dimension line.
  - The two origins of the dimension line.
  - The dimension text.
- You can change the position of the text by clicking its grip and moving it parallel to the dimension line.
- You can change the position of the dimension line by clicking one of the two grips and moving it closer to, or farther from, the origin.
- You can change the measured distance by moving one of the two grips of the origin, so the distance will change accordingly.
- See the example below:

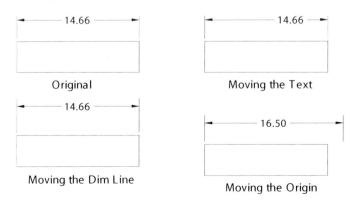

- Also, if you select a dimension block and right-click, the following shortcut menu will appear (the part concerning the dimension only is shown below):

- You can change four things in the selected dimension block.

Dim Text Position
- You can change the position of the dimension text with the following options:
  - **Above dim line**, it will change it from any other postion to Above.
  - **Centered**, it will change it from any other postion to Centered.
  - **Home text**, it will restore the postion to its origianl position according to its dimension style.
  - **Move text alone**, it give the user the ability to position the text anywhere.
  - **Move with leader**, to do the same as above but with a leader.
  - **Move with dim line**, you move both the dimension text and the dimension line in one single step.

Precision
- This will allow the user to set the number of decimal places for the number shown.
- Users can start from no decimal places up to 8 decimal places.

Dim Style
- The changes made using this method can be saved in a new dimension style. Use the option **Save as New Style**, and type in a new name.
- Users can change the dimension style of any dimension block in the drawing. The existing dimension styles will appear, select the new desired dimension style.

Flip Arrow
- To flip the arrows from inside to outside and vice versa.
- The process is done only on one of the arrows and not two.

DIMENSION BLOCK PROPERTIES

- Select a dimension block, then right-click. A shortcut menu will appear. The last option will be **Properties**. The following dialog box will appear:

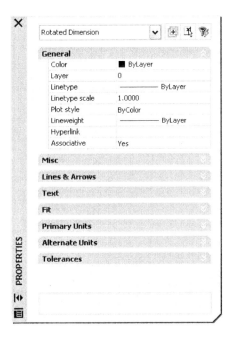

FIGURE 9-13

- Under **General**, you will see the general properties of the selected dimension block.
- Then you will see Misc, Lines & Arrows, Text, Fit, Primary Units, Alternate Units, and Tolerances. If you comapre these to the Dimension Style command that we discussed, you will find them identical. Which means, you can change any of the characteristics of the dimension block after we place it through **Properties**.

PUTTING DIMENSIONS ON THE PLAN (METRIC)

 Workshop 18-A

1. Start AutoCAD 2008, and open the file: **Small_Villa_Ground_Floor_Plan_Metric_Workshop_18.dwg**. (If you solved the previous workshop correctly, open your file.)
2. Make the **Dimension** layer the current layer.
3. Make the following layers frozen: **Furniture**, **Hatch**, and **Text**.

4. Using the **Outside** and **Inside** dimension styles, put the dimensions for the outer and inner dimensions as shown:

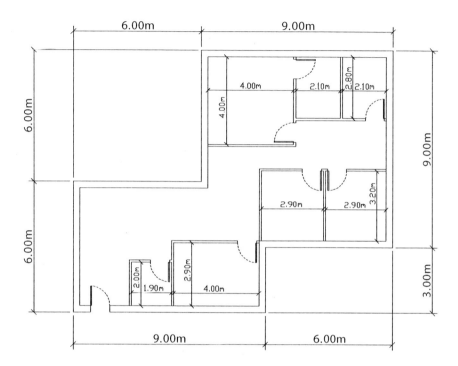

5. Save the file and close it.

PUTTING DIMENSIONS ON THE PLAN (IMPERIAL)

Workshop 18-B

1. Start AutoCAD 2008, and open the file: **Small_Villa_Ground_Floor_Plan_Imperial_Workshop_18.dwg**. (If you solved the previous workshop correctly, open your file.)
2. Make the **Dimension** layer the current layer.
3. Make the following layers frozen: **Furniture**, **Hatch**, and **Text**.

4. Using the **Outside** and **Inside** dimension styles, put the dimensions for the outer and inner dimensions as shown:

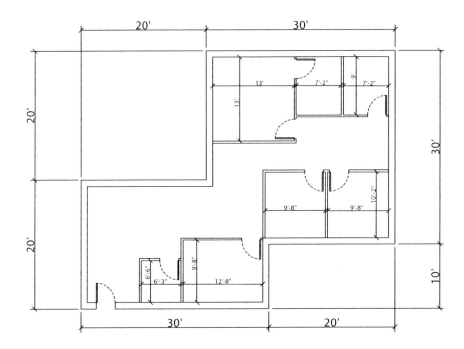

5. Save the file and close it.

NOTES

CHAPTER REVIEW

1. You can ONLY create dimension styles that will affect all dimension types.
   a. True
   b. False
2. _____ and _____ are two types of dimensions that you can use with arcs.
3. One of the following is not an AutoCAD dimension command:
   a. dimlinear
   b. dimarc
   c. dimchordlength
   c. dimaligned
4. The only way to put an Angular dimension is to have two lines.
   a. True
   b. False
5. Which of these is a type of Tolerance in AutoCAD:
   a. Deviation
   b. Symmetrical
   c. Limits
   d. All of the above
6. In order to make a dimension style _____ double-click the name in the dimension style dialog box.

CHAPTER REVIEW ANSWERS

1. b
2. Arc length, Jogged, Radius, Diameter
3. c
4. b
5. d
6. current

Chapter 10
PLOTTING YOUR DRAWING

In This Chapter

- Model Space vs. Paper Space
- What is a layout and how to create it?
- What is Page Setup and how to create it?
- Creating viewports in layout using multiple methods
- Editing and scaling viewports
- What is Plot styles and how to create them?
- Plot command
- DWF plotting

INTRODUCTION

- Before AutoCAD 2000, almost all users of AutoCAD plotted from Model Space, which is the place they made their design in.
- But, in AutoCAD 2000, the emergence of the new concept of **Layouts** made it easy for everybody to shift their attention to the new method, which encompasses lots of new features, and surpasses plotting from Model Space.
- Also in AutoCAD 2000, a new feature was introduced called Plot Style, which allowed the user to create color-independent configuration plotting.
- No doubt, AutoCAD 2000 was a flagship version in more than one aspect, but the new improvements in plotting process made it the most important.

MODEL SPACE VS. PAPER SPACE

- Model Space is the place where users will create their drawing, with all of the modification process.

- When users start thinking about plotting, Paper Space should be the one to use.
- There is only one Model Space in each drawing file.
- Before AutoCAD 2000, there was only one Paper Space per drawing file.
- From AutoCAD 2000 and on (AutoCAD 2000i, 2002, 2004, 2005, 2006, 2007, and 2008) the name of Paper Space changed to Layout.
- You can create as many layouts as you wish in each drawing file.
- Each Layout is connected to Page Setup in which the user should specify a minimum of three things. They are:
  - Plotter you will send the drawing to.
  - Paper size you will use.
  - Paper orientation (Portrait or Landscape).
- To show the importance of this feature, let's give the following example: say we have a company who owns A0 plotter, A2 printer, along with A4 laser printer. The staff will use all of these printers to print a single drawing:
  - If you use Model Space, you will change the setup of the printer, paper size, and paper orientation each and every time you want to print.
  - But if you create three Layouts with the proper Page Setup, each time you want to print simply go to the Layout and give the command **Print**, which will save time, effort, and money!

INTRODUCTION TO LAYOUTS

- Each layout consists of the following elements:
  - **Page Setup**, for which you will specify the printer (or plotter), the paper size, paper orientation, and other things that will be covered later in this chapter.
  - **Objects**, like blocks (e.g., the title block), text, dimensions, and any other desired object.
  - **Viewports**, which will be covered separately in the coming discussion.
- Each Layout should have a name. AutoCAD will give it a temporary name, but you can change it as you wish.
- By default, the when you create a new drawing using the *acad.dwt* template, two layouts, namely Layout1 and Layout2, will be there automatically for the user.
- You can make the setting of the **Page Setup Manager** dialog box to appear when you click on a layout for the first time for the purpose of setting the printer, paper size, etc.

- **Layouts** and **Page Setup** will be saved in the drawing file.
- By default, **Page Setup** has no name, but the user can name it and use it in other Layouts in the current drawing file or in other drawings.

WHERE DO I FIND LAYOUTS?

- By default the layout tabs are found at the lower left corner of the screen.
- If you don't see them, do the following:
  - From menus select **Tools/Options**. The Option dialog box will appear.
  - Select the **Display** tab under **Layout elements**. Make sure that the **Display Layout and Model tabs** option is checked on.

HOW TO CREATE A NEW LAYOUT

- There are several ways to create a new layout, and these are:

Using Right-click
- A very simple method. Right-click on any existing layout and a shortcut menu will appear. Select a **New Layout** option.
- A new layout will be added with a temporary name. You can rename using the right-click and selecting the **Rename** option.

Using a Template
- You can bring any layout defined inside a template and use it in your current drawing file.
- Right-click on any existing layout a shortcut menu will appear. Select the **From template** option and the following dialog box will appear:

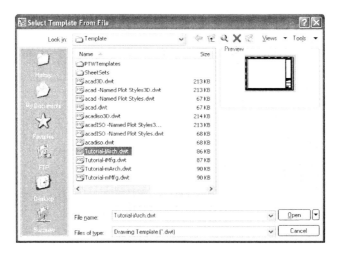

FIGURE 10-1

- Select the desired template and click **Open**. The following dialog box will appear:

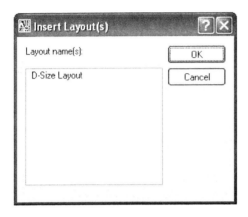

FIGURE 10-2

- Click on one of the listed layouts and click **OK**.

Move or Copy
- Using this option you can move a layout from its current position to the left or to the right. Also, you can create a copy of an existing layout.
- Select the desired layout that you want to create a copy from and right-click. A shortcut menu will appear. Select the **Move or Copy** option and the following dialog box will appear:

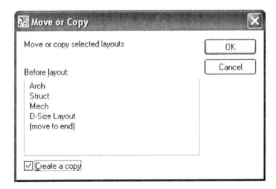

FIGURE 10-3

- You can move the layout position relative to the other layouts by clicking the layout name, holding, and dragging it to the position required.
- From the upper dialog box you can see there are four existing layouts, select one of them and click the check box **Create a copy** on. The layout you select in this list will be at the right of the new layout.

Using Layout Command

- This command will create a new layout and place it as the last layout on the right.
- To issue this command use one of the following ways:
  - From the **Layout** toolbar, click the **New Layout** button.
  - From menus select **Insert/Layout/New Layout**.
  - Type **layout** in the Command window.
- The following prompt will appear:

```
Enter new Layout name <Layout1>: (Type in the name
of the new layout)
```

- A new layout will be added.

Layout Wizard

- To create a layout using a wizard.
- To issue this command use one of the following two methods:
  - From menus select **Insert/Layout/Create Layout Wizard**.
  - Type **layoutwizard** in the Command window.
- Either way you will see the following dialog box:

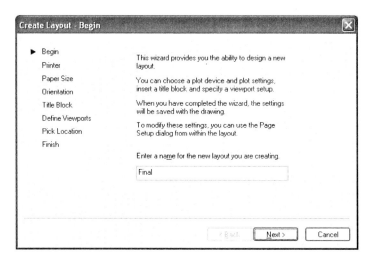

FIGURE 10-4

- Type in the name of the new layout, and click **Next**. The following dialog box will appear:

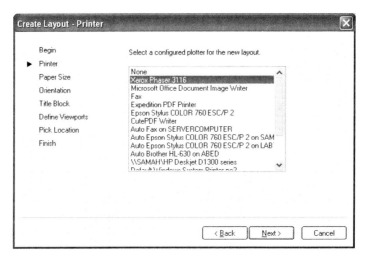

FIGURE 10-5

- Select the desired printer and click **Next**. The following dialog box will appear:

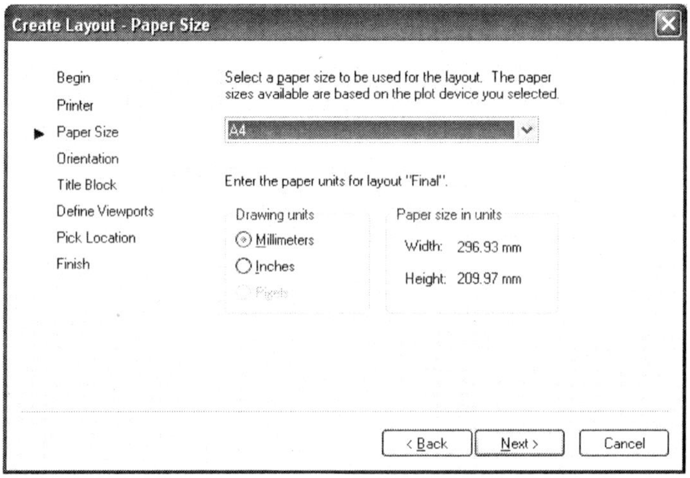

FIGURE 10-6

- Select the desired paper size and units, and click **Next**. The following dialog box will appear:

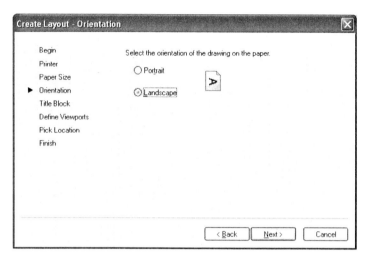

FIGURE 10-7

- Select the paper orientation (Portrait or Landscape) and click **Next**. The following dialog box will appear:

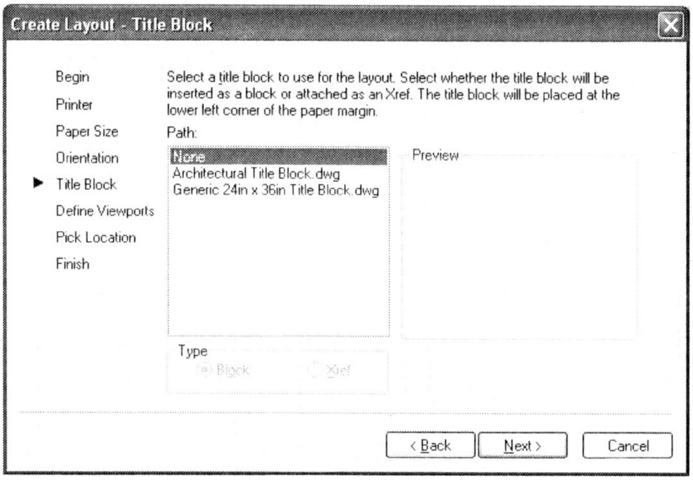

FIGURE 10-8

- Select a predefined title block (unless you prefer to use your own) or None, and click **Next**. The following dialog box will appear:

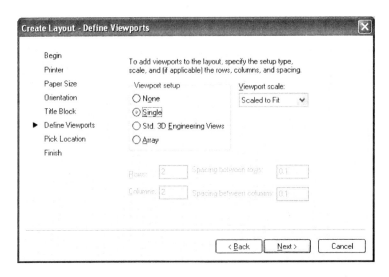

FIGURE 10-9

- Select the viewport setup (this can be done in a separate step) or None, and click **Next**. The following dialog box will appear:

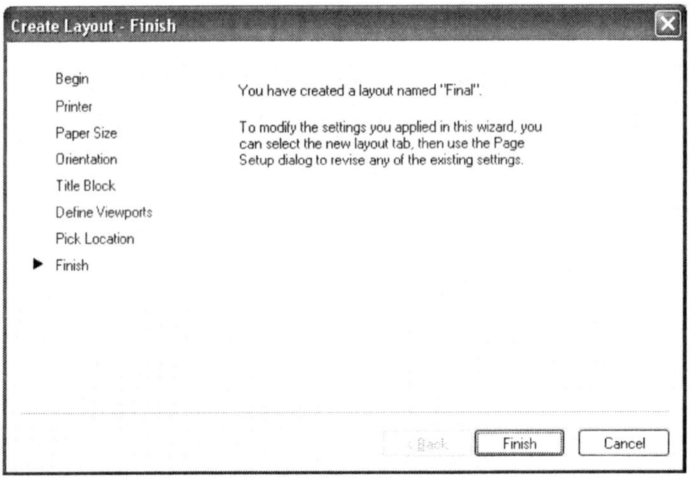

FIGURE 10-10

- Click **Finish**. Now you have a new layout as per your requirements.
- This method will automatically create a Page Setup that will hold the same name as the layout.

WHAT IS PAGE SETUP MANAGER?

- As we said previously, each layout will have a Page Setup linked to it.
- The Page Setup Manager is the dialog box in which you will create, modify, delete, and import Page setups in the Model or Paper Space.
- To issue this command use one of the following methods:
  - From the **Layout** toolbar, click the **Page Setup Manager** button.
  - From menus select **File/Page Setup Manager**.
  - Type **pagesetup** in the Command window.
- The following dialog box will appear:

FIGURE 10-11

- At the top, users will see the **Current layout** name and at the bottom the **Selected page setup details**.

- A check box will allow users to **Display Page Setup Manager when creating a new layout**.

- To create a new Page Setup, click the **New** button. The following dialog box will appear:

FIGURE 10-12

- Type in the name of the new Page Setup and click **OK**. The following dialog box will appear:

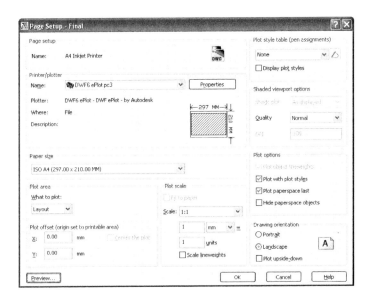

FIGURE 10-13

- Specify the **Name** of the printer or the plotter you want to use (this printer should be installed and configured beforehand).
- Specify what **Page Size** should be used.
- Specify **What to plot**. You have three choices: **Display**, **Layout**, and **Window**. If you use Paper Space (as this book recommends) always leave it as **Layout**.
- Specify the **Plot Offset**. By default it is 0,0 and it starts from the printable area (the truncated paper size, which is the original paper size minus the printer margins).
- Specify the **Plot Scale**. If you want to plot from Layout, then you use the viewports (which will be discussed shortly), and you will specify plot scale for each viewport. Accordingly, you set this Plot scale to 1=1. Specify if you want to **scale lineweights** or not.
- Specify the **Plot style table (pen assignment)**. This will be discussed later in this Chapter. Specify whether to **Display** the effects of the **plot style** on the layout or not.
- If you are plotting a 3D drawing and you want to plot it as shaded or rendered, then specify the **Quality** of the shading or rendering.
- Specify the **Plot options**, which are:
  - Specify to plot the objects with their lineweight as specified for each object and layer. This will be available only if you specify **None** for the **Plot style** setting.
  - Specify to let the Plot Style control the objects and layers lineweight.
  - By default Paper Space objects will be printed first, and then the Model Space objects. Specify if you want the opposite.
  - Specify to show or hide the Paper Space objects.
- Specify the **Paper orientation**, whether **Portrait** or **Landscape**. By default the printer will start printing from top-to-bottom. Specify if you want the opposite.
- When you are done, click **OK**. The Page Setup you created will be used in the current layout and any layout in the current drawing file.
- To link any layout in your drawing file to a certain Page Setup, go to the desired layout, start the **Page Setup Manager**, select the Page Setup from the list, then click **Set Current** (also, you can double-click the name of the Page Setup). Now the current layout is linked to the Page Setup you select.
- To modify the settings of an existing Page Setup.
- To bring a saved Page Setup from an existing file.

CREATING LAYOUTS AND PAGE SETUP (METRIC)

 Workshop 19-A

1. Start AutoCAD 2008, and open the file: **Small_Villa_Ground_Floor_Plan_Metric_Workshop_19.dwg**. (If you solved the previous workshop correctly, open your file.)
2. Make sure the current layer is **Viewports**.
3. Right-click on the name of any existing layout and select **From template**.
4. Select the template file **Tutorial-mArch.dwt**.
5. Select the layout name **ISO A1 Layout**.
6. Go to **ISO A1 Layout** and delete the only viewport in the layout (select its frame and press the [Del] key).
7. Select **Insert/Layout/Create Layout Wizard** and set the following:
   a. Name = **Final**
   b. Printer = **DWF6 ePlot.pc3**
   c. Paper Size = **ISO A3 (420 x 297 MM)**
   d. Drawing Units = **MM**
   e. Orientation = **Landscape**
   f. Title Block = **None**
   g. Viewports = **None**
   h. **Finish**
8. Make the layer **Frame** current and insert the file **ISO A3 Landscape Title Block.dwg** using 0,0 as insertion point.
9. Save the file and close it.

CREATING LAYOUTS AND PAGE SETUP (IMPERIAL)

 Workshop 19-B

1. Start AutoCAD 2008, and open the file: **Small_Villa_Ground_Floor_Plan_Imperial_Workshop_19.dwg**. (If you solved the previous workshop correctly, open your file.)
2. Make sure the current layer is **Viewports**.
3. Right-click on the name of any existing layout and select **From template**.
4. Select the template file **Tutorial-iArch.dwt**.
5. Select the layout name **D-Size Layout**.
6. Go to **D-Size Layout** and delete the only viewport in the layout (select its frame and press the [Del] key).

7. Select **Insert/Layout/Create Layout Wizard** and set the following:
   a. Name = **Final**
   b. Printer = **DWF6 ePlot.pc3**
   c. Paper Size = **ANSI B (17 x 11 Inches)**
   d. Drawing Units = **Inches**
   e. Orientation = **Landscape**
   f. Title Block = **None**
   g. Viewports = **None**
   h. **Finish**
8. Make the **Frame** layer current and insert the file **ANSI B Landscape Title Block.dwg** using 0,0 as insertion point.
9. Save the file and close it.

LAYOUTS AND VIEWPORTS

- You will see the following after you complete:
    - Creating a new layout.
    - Creating a Page Setup.
    - Linking a Page Setup to the layout.

- The outer frame is the real paper size.

- The inner frame (dashed) is the truncated paper size, which is the paper size minus the margins of the printer.
  - Each printer comes from the manufacturer with built-in margins from all sides.
  - AutoCAD can read these margins from the printer driver.
  - This means that each user should read the manual of the printer in order to know exactly how much these margins are on each side.
  - This will prove vital when users want to create the frame block of the company or establishment they work in because users should create it within the truncated paper size and not the full size.
- The above proves that printing from Paper Space is WYSIWYG (What You See Is What You Get).
- Also, by default you will see that a single viewport of your drawing appears at the center of the paper size.
- As we said in the beginning of this chapter, we have only one Model Space, yet we can have as many Layouts as we wish. **Viewport** is a rectangular shape (or any irregular shape) that contains a view of the Model Space.
- There are two types of Viewports:
  - Model Space Viewports, which are tiled always, cannot be scaled, and the arrangement of viewports shown on the screen cannot be printed.
  - Paper Space Viewport, which can be tiled or separated, can be scaled, and the arrangement of viewports shown on the screen can be printed.
- Check the following illustration:

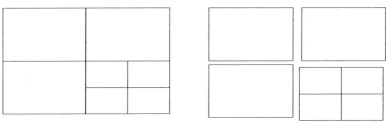

Model Space Viewports   Paper Space Viewports

ADDING VIEWPORTS

- Users can add viewports to layouts using different methods:
  - Adding a single rectangular viewport.
  - Adding multiple rectangular viewports.
  - Adding a single polygonal viewport.

- Converting an object to be a viewport.
- Clipping an existing viewport.
- In the coming pages we will discuss each method.

Single Rectangular Viewport

- You can add several single rectangular viewports in any layout you want, and in as many as you wish. You have to specify two opposite corners in order to specify the area of the viewport.
- To issue this command use one of the following ways:
  - From the **Viewports** toolbar, click **Single Viewport**.
  - From menus select **View/Viewports/1 Viewport**.
  - Type **-vports** in the Command window.
- The following prompt will appear:

```
Specify corner of viewport or
[ON/OFF/Fit/Shadeplot/Lock/Object/Polygonal/Restore
LAyer/2/3/4] <Fit>: (Specify the first corner)
Specify opposite corner: (Specify the opposite
corner)
```

- Check the example below:

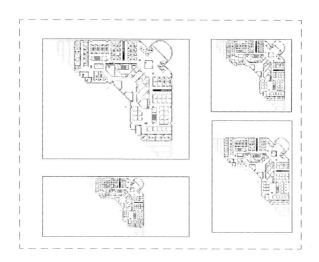

Multiple Rectangular Viewport

- You can add multiple rectangular viewports in a single command, but each viewport is considered a single object.
- To issue this command use one of the following methods:
  - From the **Viewports** toolbar, click the **Display Viewport Dialog** button.

- From menus select **View/Viewports/New Viewports**.
- Type **vports** in the Command window.
  - The following dialog box will appear:

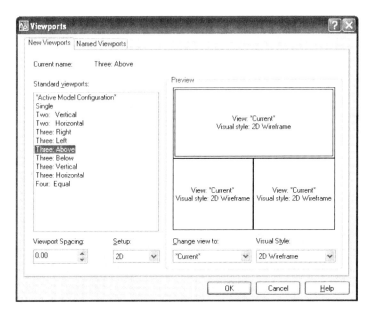

FIGURE 10-14

  - Specify the arrangement you like. You can have two (Horizontal or Vertical) or six different arrangements for the three viewports, and one for the four viewports.
  - If you want the viewports to be tiled, leave the **Viewport Spacing** = 0, otherwise set a new value.
  - Click **OK**. AutoCAD will display the following prompt:

```
Specify first corner or [Fit] <Fit>: (Specify the
first corner)
Specify opposite corner: (Specify the opposite
corner)
```

  - Check the illustration on the next page.
  - Example of three (Above) with spacing between the viewports:

PLOTTING YOUR DRAWING   271

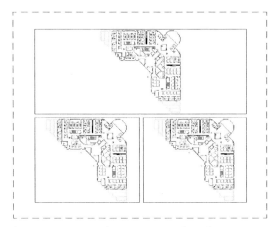

Single Polygonal Viewport

- To add a single viewport with any irregular shape consisting of both straight lines and arcs.
- To issue this command use one of the following methods:
  - From the **Viewports** toolbar, click the **Polygonal Viewport** button.
  - From menus select **View/Viewports/Polygonal Viewport**.
  - Type **–vports** in the Command window, then type **p**.
- The following prompt will appear:

```
Specify start point:
Specify next point or [Arc/Length/Undo]:
Specify next point or [Arc/Close/Length/Undo]:
```

- It is almost identical to the **Pline** command.
- Check the illustration on the next page.
- Example of a polygonal viewport:

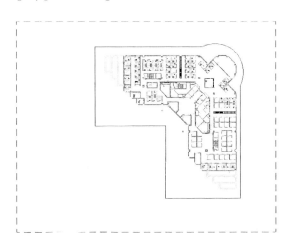

Converting an Object to Viewport

- To convert an existing object to viewport.
- To issue this command use one of the following methods:
  - From the **Viewports** toolbar, click the **Convert Object to Viewport** button.
  - From menus select **View/Viewports/Object**.
  - Type **–vport** in the Command window, then type **o**.
- The following prompt will appear:

```
Select object to clip viewport: (Select the desired
object)
```

- Check the illustration below (circle and pline converted to viewports):

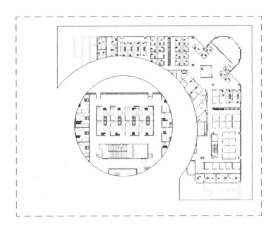

Clipping an Existing Viewport

- If you have a rectangular viewport you can change it to an irregular shape by clipping it.
- To issue this command use one of the following methods:
  - From the **Viewports** toolbar, click the **Clip existing viewport** button.
  - Type **vpclip** in the Command window.
- The following prompt will appear:

```
Select viewport to clip:
Select clipping object or [Polygonal] <Polygonal>:
Specify start point:
Specify next point or [Arc/Length/Undo]:
Specify next point or [Arc/Close/Length/Undo]:
```

- First select the existing viewport. You can draw a pline beforehand, or you can draw any irregular shape using the **Polygonal** option (which is identical to Polygonal viewport).

- Check the illustration below. Before clipping:

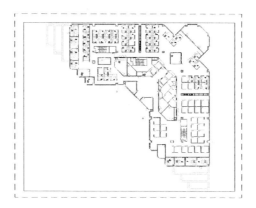

- After clipping:

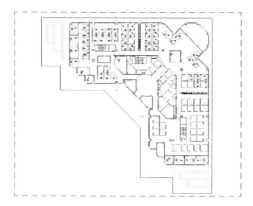

MODEL SPACE AND PAPER SPACE MODES IN LAYOUTS

- In a layout you deal with the viewports in two modes:
  - Paper Space mode.
  - Model Space mode.

Paper Space
Mode
- This is the default mode in any layout.
- In the **Status** bar you will see the following:

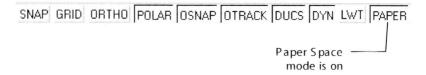

Paper Space mode is on

- In this mode you can place the viewports, as we learned in the previous discussion.
- Also, you can deal with the viewports as objects, hence you can copy, move, stretch, and delete them.

Model Space Mode

- In this mode you will get *inside* the viewport.
- You can zoom in, zoom out, and pan while you are in this mode.
- Also, you can scale each viewport.
- Moreover, you can change the status of layers for the current viewports.
- There are two ways to be in this mode:
  - Double-click inside the desired viewport.
  - From the Status bar, click the **Paper** button and it will switch to **Model** as shown below:

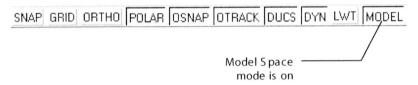

Model Space mode is on

 - In order to switch from Model Space mode to Paper Space mode, simply double-click outside any viewport.

MODIFYING, SCALING, AND MAXIMIZING VIEWPORTS

- Each viewport can be modified, scaled, and maximized to fill the whole screen.

Modifying
- Each viewport is considered as an object, hence it can be copied, moved, scaled, and deleted. You have to select each viewport from its *border* in order to select it.
- You can select viewports first, and then issue the modifying commands, or vice versa.

Scaling
- Each viewport can be scaled relative to the Model Space units.
- Double-click inside the desired viewport. You will switch to the Model Space mode for this viewport.
- From the **Viewports** toolbar select the scale pop-up list and select the desired scale, just like the example below:

Scale pop-up list

- If you didn't find the desired scale, simply write it down in the pop-up list like 1:14 and AutoCAD will accept it from you.
- After you set the scale you can use the **Pan** command, but if you want to use the **Zoom** command, the scale value will be invalid, hence you have to repeat the procedure of setting the scale again.
- In order to avoid this problem you can lock the display of the viewport. Follow these steps:
  - Make sure you are in the Paper Space mode.
  - Select the desired viewport.
  - Right-click and a shortcut menu will appear. Select **Display Locked**, then select **Yes**.
  - After this step, even if you were in the Model Space mode, and you start the **Zoom** command, the whole layout will zoom and not just the selected viewport.
- *NOTE* After you make the scaling of a viewport there are two possible results:
  - The scale is perfect to the area of the viewport. Leave it as is.
  - The scale is either too small or too big, so either change the scale or change the area of the viewport.

Maximizing
- After placing your viewports, there will be small ones and big ones.
- For small ones you can maximize the area of the viewport to be as large as your screen momentarily. You do all of your work needed, and then get it back to the original size.
- From the **Status** bar, click the **Maximize Viewport** button:

- The same button will be changed to **Minimize button** in order to restore the original size of the viewport.
- *NOTE* Another way of Maximizing viewport is to double-click the *border* of this viewport.

FREEZING LAYERS IN VIEWPORT

- We learned in a previous chapter how to freeze a layer. This tool will be effective in both Model Space and Layouts.
- In Layouts, if you freeze a layer it will be frozen in all viewports. So, what if you want to freeze a certain layer(s) in one of the viewports and not the other viewports. To do so you have to freeze the layer in the current viewport.

- Do the following:
  - Make the desired viewport current (by double-clicking inside it).
  - From the **Layers** toolbar or from **Layer Properties Manager**, click the icon **Freeze or thaw in current viewport** for the desired layer, just like the illustration below:

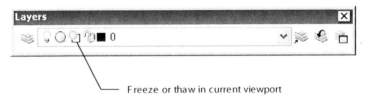

Freeze or thaw in current viewport

INSERTING AND SCALING VIEWPORTS (METRIC)

 Workshop 20-A

1. Start AutoCAD 2008, and open the file: **Small_Villa_Ground_Floor_Plan_Metric_Workshop_20.dwg**. (If you solved the previous workshop correctly, open your file.)
2. Select **ISO A1 Layout**.
3. Make the **Viewports** layer current.
4. Click OSNAP off.
5. From the **Viewports** toolbar, click the **Display Viewport Dialog** button, select the arrangement **Three: Left**, and set the **Viewport spacing** to 5.
6. Click **OK** and specify two opposite corners so the three viewports fill the empty space, just like the following:

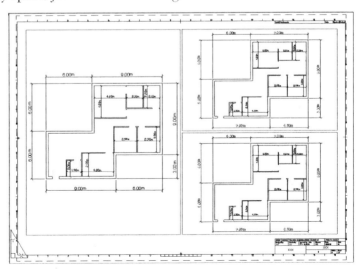

7. Select the big viewport on the left and set the scale to 1:50. The upper viewport scale to 1:20, and finally the lower viewport scale to 1:30.
8. Double-click outside the viewports to move to the **Paper** mode.
9. Freeze the **Dimension** layer and thaw **Furniture**, **Hatch**, and **Text**.
10. Make the big left viewport the current viewport and freeze only in this viewport the layers **Furniture**, **Hatch**, and **Text**.
11. Make the upper viewport current and pan to the Master Bedroom (don't use the zooming facilities). Make the lower viewport current and pan to the Living Room.
12. Double-click outside the viewports to move to the **Paper** mode.
13. Select the three viewports, right-click, and from the shortcut menu select **Display locked**, then **Yes**.
14. The drawing should look like the following:

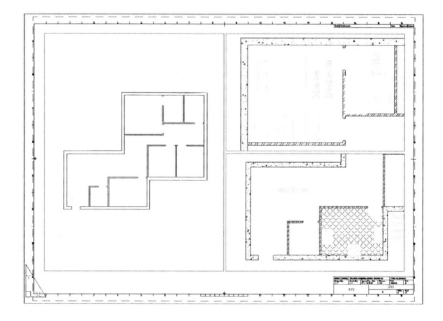

15. Save the file and close it.

INSERTING AND SCALING VIEWPORTS (IMPERIAL)

 Workshop 20-B

1. Start AutoCAD 2008, and open the file: **Small_Villa_Ground_Floor_Plan_Imperial_Workshop_20.dwg**. (If you solved the previous workshop correctly, open your file.)

2. Select **D-Sized Layout**.
3. Make the **Viewports** layer current.
4. Click OSNAP off.
5. From the **Viewports** toolbar, click the **Display Viewport Dialog** button, select the arrangement **Three: Left**, and set the **Viewport spacing** to 0.35.
6. Click **OK** and specify two opposite corners so the three viewports fill the empty space, just like the following:

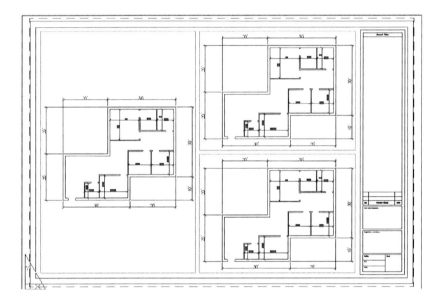

7. Select the big viewport on the left and set the scale to $1/4'' = 1'$, the upper viewport scale to $1/2'' = 1'$, and the lower viewport scale to $1/2'' = 1'$.
8. Double-click outside the viewports to move to the **Paper** mode.
9. Freeze the **Dimension** layer and thaw **Furniture**, **Hatch**, and **Text**.
10. Make the big left viewport the current viewport and freeze only in this viewport the layers **Furniture**, **Hatch**, and **Text**.
11. Make the upper viewport current and pan to the Master Bedroom (don't use the zooming facilities). Make the lower viewport current and pan to the Living Room.
12. Double-click outside the viewports to move to the **Paper** mode.
13. Select the three viewports, right-click, and from the shortcut menu select **Display locked**, then **Yes**.

14. The drawing should look like the following:

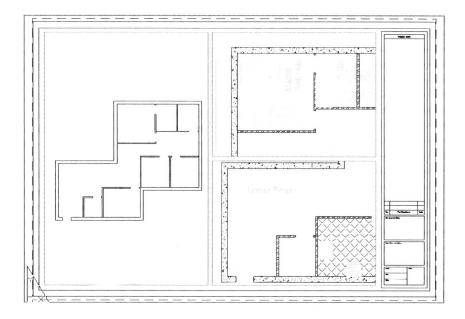

15. Save the file and close it.

PLOT STYLE TABLES: INTRODUCTION

- Users of AutoCAD may use lots of colors in the drawing file, but will these colors print out?
- There are two possibilities:
  - Either users will use the same colors in both the softcopy and the hardcopy of the drawing.
  - Or, users will assign for each color in the softcopy a different color in the hardcopy.
- To translate the colors between softcopy and hardcopy, we need to create a Plot Style.
- There are two types of Plot Styles:
  - Color-dependent Plot Style Table.
  - Named Plot Style Table.

PLOT STYLE TABLES: COLOR-DEPENDENT PLOT STYLE TABLE

- This method is almost the same method that was used prior to AutoCAD 2000. It depends on the colors used in the drawing file.
- Each color used in the drawing file will be printed with a color chosen by the user. Also, users will set the lineweight, linetype, etc., for each color.
- This method is limited because you will have only 255 colors to use.
- Also, if you have two layers with the same color you be will forced to use the same output color, lineweight, and linetype.
- Each time you create a Color-dependent Plot Style Table, AutoCAD will ask you to name the file with the extension *.*ctb*
- You can create Plot Style tables from outside AutoCAD (using the Control Panel of Windows), or from inside AutoCAD using the Wizards. This will only initiate the command, but the command is the same for both ways.
- From outside AutoCAD, start the Control Panel of Windows, double-click **Autodesk Plot Style Manager** icon, then double-click **Add-A-Plot Style Table Wizard**.
- Inside AutoCAD, from menus **Tools/Wizards/Add Plot Style Table**.
- Either way the following dialog box will appear:

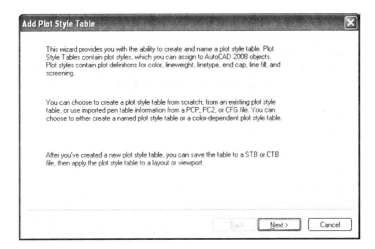

FIGURE 10-15

- In the above dialog box, AutoCAD explains the next few steps to take. Click **Next** and the following dialog box will appear:

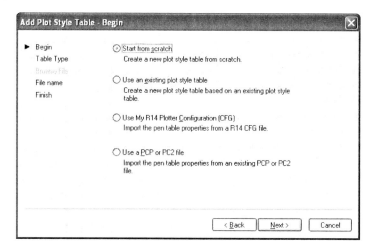

FIGURE 10-16

- You have four choices to select from. They are:
  - Start creating your style from scratch.
  - Use an existing plot style.
  - Import the AutoCAD R14 CFG file and create a plot style from it.
  - Import a PCP or PC2 file and create a plot style from it.
- Select **Start from scratch** and click **Next**. The following dialog box will appear:

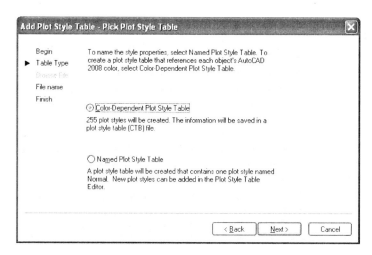

FIGURE 10-17

- Select **Color-Dependent Plot Style Table** and click **Next**.

- The following dialog box will appear:

FIGURE 10-18

- Type in the name of the plot style and click **Next**. The following will appear:

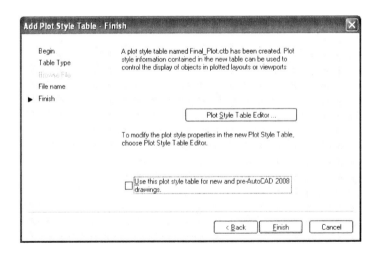

FIGURE 10-19

- You can **Use this plot style table for new and pre-AutoCAD 2008 drawings**. Click the **Plot Style Table Editor** button and you will see the following:

FIGURE 10-20

- At the left part, select the color you used in your drawing file, then at the right part, change all or any of the following settings:
  - **Color**, which is the hardcopy color.
  - **Dither**, this option will be dimmed if your printer or plotter doesn't support Dithering. Dither means, to allow the printer to give the impression of using more colors than the limited 255 colors of AutoCAD. It is preferable to leave this option off. It should be on, if you want **Screening** to work.
  - **Grayscale**, you can translate the 255 colors to grayscale grades (good for laser printer printing).
  - **Pen #**, good only for the old types of pen plotters, which are nowadays obsolete.
  - **Virtual pen #**, for nonpen plotters to simulate pen plotters by assigning a virtual pen for each color. It is preferable to leave it **Automatic**.
  - **Screening**, this is good for trial printing, which if you give numbers less than 100, will reduce the intensity of the shading and fill hatches. You should turn **Dithering** on so **Screening** will be effective.

- **Linetype**, you can use the object's linetype, or you can set a different linetype for each color.
- **Adaptive**, to change the linetype scale to fit the current line length, so it will start with a segment and end with a segment, and not end with a space. Turn this option off if the linetype scale is important to the drawing.
- **Lineweight**, set the lineweight for the color selected. You have a list of lineweights to select from.
- **Line end style**, to specify the end style for lines. The available end styles are: Butt, Square, Round, and Diamond.
- **Line join style**, to specify the line join (the connection between two lines) style. The available join styles are: Miter, Bevel, Round, and Diamond.
- **Fill style**, to set the fill style for an area filled in the drawing (good for trial printing).
- Click **Save & Close**. Then click **Finish**.
- To link Color-dependent Plot Style to a layout, do the following steps:
  - Go to the desired Layout, then start the **Page Setup Manager**.
  - At the upper right part of the dialog box, change the **Plot style table (pen assignment)** setting to the desired **ctb** file:

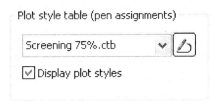

  - Click the checkbox **Display plot styles** on.
- You can assign for each layout one **ctb** file.
- In order to see the lineweight of the objects you have to switch the **LWT** button on the Status bar on.

PLOT STYLE TABLES: NAMED PLOT STYLE TABLE

- This method is the new method introduced for the first time in AutoCAD 2000. It does not depend on colors.
- You will create a Plot Style and give it a name. Each plot style will include inside of it different tables, which you will link later on with layers.

- With this method you can have two layers having the same color but will be printed in different colors, linetypes, and lineweights.
- The Named Plot Style Table has a file extension of *.*stb*.
- The creation procedure of Named Plot Style is identical to Color-dependent Plot Style, except the last step, which is the **Plot Style Table Editor** button.
- From outside AutoCAD, start Control Panel, double-click the **Autodesk Plot Style Manager** icon, then double-click **Add-A-Plot Style Table Wizard**.
- From Inside AutoCAD, select from menus **Tools/Wizards/Add Plot Style Table**.
- Go through the dialog boxes until you reach the **Plot Style Table Editor** button. Click it and you will see the dialog box on the next page.
- Plot Style Table Editor for Named Plot Style, after we clicked the **Add Style** button:

FIGURE 10-21

- As you can see, you can change all or any of the following:
  - Type in the Name of the style.
  - Type in any Description for this style.

- Specify the Color that you will use in the hardcopy.
- The rest is identical to Color-dependent Plot Style Table.
▪ You can add as many styles as you wish in the same Named Plot Style.
▪ Click **Save & Close**. Then click **Finish**.
▪ In order to link a Named Plot Style Table with any drawing, do the following steps:
    - You have to first convert one of the **.ctb** files to a **.stb** file. In the Command window type **convertctb**, and a dialog box with all the .ctb files will appear. Select one of them, keeping the same name, or giving a new name, then click **OK**.
    - Convert the drawing from Color-dependent Plot Style to Named Plot Style. In the Command window type **convertpstyles**, and the following warning message will appear:

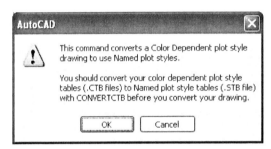

FIGURE 10-22

- Click OK. The following dialog box will appear:

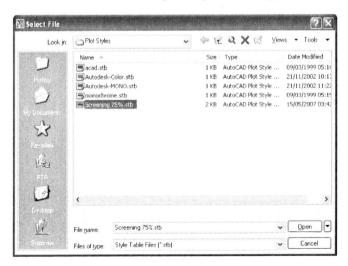

FIGURE 10-23

- Select the Named Plot Style Table you just converted, and click Open. The following message will appear in the Command window:

Drawing converted from Color-dependent mode to Named plot style mode.

- Because users will use the Named Plot Style Table with layers, the Model Space and all layouts will be assigned the same .stb file.
- Go to the desired Layout, then start the Page Setup Manager. At the upper right part of the dialog box, change the Plot style table (pen assignment) setting to the desired .stb file. Click the checkbox Display plot styles on and end the Page Setup Manager command.

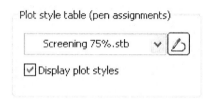

- Go to the **Layer Properties Manager**, and for a certain layer under the **Plot Style** column, click **Normal**. The following dialog box will appear. Select the name of the Plot Style desired, and click **OK**.

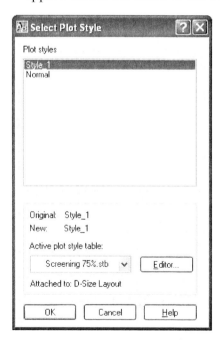

FIGURE 10-24

PLOT STYLE TABLES

Exercise 32

1. Start AutoCAD 2008.
2. Open the file **Exercise_32.dwg**.
3. Click Layout1. When the dialog box appears, click Close (this means we are accepting the default Page Setup to this layout).
4. Start a new Plot Style Table and choose Color-dependant. Call this Plot Style **Mechanical_BW** and make the following changes:

Drawing Color	Plotter Color	Linetype	Lineweight
Color 2	Black	Dashdot	0.30
Color 3	Black	Solid	0.70
Color 4	Black	Solid	0.50
Color 6	Black	Solid	0.50

5. Go to the Page Setup Manager, select **Plot style table** to be **Mechanical_BW**, and make **Display plot styles** on.
6. Click **LWT** to see the effect of the lineweight.
7. Save As the file **Exercise_32_1.dwg**.
8. In the Command window, type **convertctb**, and convert **Mechanical_BW** from a **.ctb** to a **.stb** file.
9. In the Command window, type **convertpstyles** to convert the whole drawing to accept Named Plot Styles. Select **Mechanical_BW** when you are asked to select an **.stb** file.
10. Start a new Plot Style Table and choose Name Plot Style. Call this Plot Style **Design_Process**, and make the following changes:

Style	Description	Color	Linetype	Lineweight
Finished	Design is Final	Black	Solid	0.7
Incomplete	Incomplete Design	Green	Dashed	0.3

11. Go to the Page Setup Manager, select **Plot style table** to be Design_Process, and make **Display plot styles** on.
12. Start the Layer Manager and set the plot style for layer **Base** to **Finished**, and layers **Shaft** and **Body** to **Incomplete**.

13. Click **LWT** to see the effect of the lineweight.
14. Save the file and close it.

PLOT COMMAND

- The final step in this process is to issue the command of Plot, which will send your layout to the printer or plotter.
- As a first step, go to the desired layout that you want to plot.
- To issue this command use one of the following methods:
  - From the **Standard Annotation** toolbar, click the **Plot** button.
  - From menus select **File/Plot**.
  - Type **plot** in the Command window.
  - Right-click on the layout tab and from the shortcut menu select **Plot** option.
- The Plot dialog box:

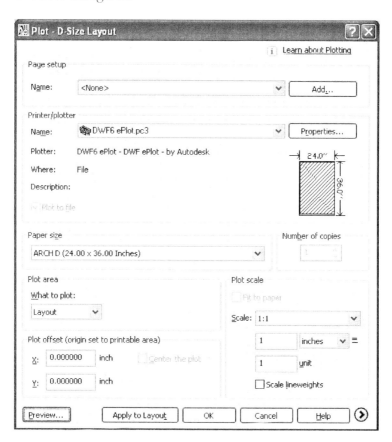

FIGURE 10-25

- As you can see, all the settings are identical to the Page setup settings.
- If you change any of these settings, AutoCAD will detach the Page setup from the current layout.
- Click the **Apply to Layout** button if you want this Plot dialog box saved with this layout for future use.
- Click **Preview** button in order to see the final printed drawing on the screen before the real printout, so you can decide if your choices of Plot styles and the other settings are correct or not.
- This button is equal to:

  - From the **Standard** toolbar, click **Print Preview**.
  - From menus select **File/Print Preview**.
  - Type **preview** in the Command window.
- After you are done, click **OK**, so the drawing will be sent to the printer.

WHAT ARE DWF FILES?

- Assume one or all of the following cases:
  - You want to share your design with another company, but you are afraid if you send them the **.dwg** file they will mess up with it.
  - Your **.dwg** file is very large (more than 1 Mb), which may not be accepted by your email engine.
  - The recipient doesn't have AutoCAD to view the **.dwg** file.
- So you don't want the other people to receive your **.dwg** file, but they have to receive your design. How can you solve such a predicament?
- AutoCAD offers you to plot to a DWF file (**D**esign **W**eb **F**ormat). This file has the following features:
  - Does not need AutoCAD to open it. Instead, free software comes with AutoCAD called **Autodesk DWF Viewer** (which you can download from Internet free of charge).
  - You can view the file, zoom, pan, and print it.
  - Small size so you can send it via email.
  - The recipient can't modify it.

HOW CAN WE PRODUCE SINGLE-SHEET AND MULTIPLE-SHEET DWFS?

Single-page DWF
Will produce a DWF file that contains a single layout. Do the following:
- At the **Page Setup Manager** select the printer to be **DWF6 ePlot.pc3**.
- Link this **Page Setup** with the layout you want to print, and click **OK**. Once you plot this layout the following dialog box will appear:

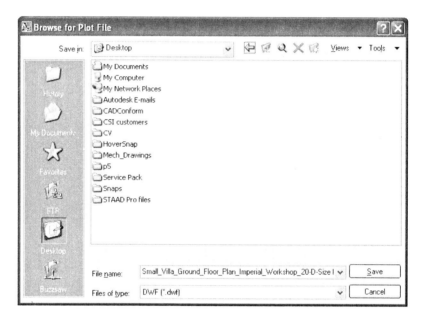

FIGURE 10-26

- Select the place in your computer to save in and type the name of the **.dwf** file. Click the **Save** button.
- In order to see this file simply double-click it and Windows will launch the Autodesk DWF Viewer automatically.

- An example of a single-sheet DWF file:

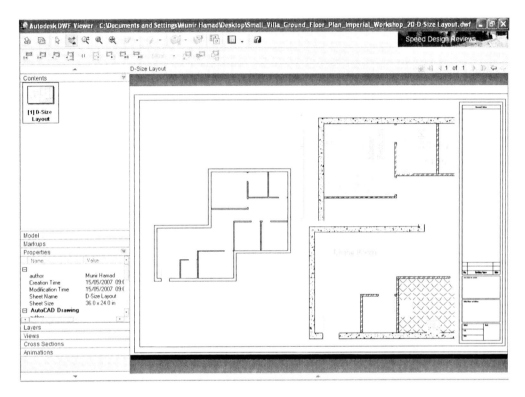

FIGURE 10-27

Multiple-page DWF

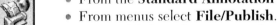

- This method will produce a DWF file that contains multiple layouts from the current drawing and from other drawings.
- Issue the **Publish** command using one of the following methods:
  - From the **Standard Annotation** toolbar, click the **Publish** button.
  - From menus select **File/Publish**.
  - Type **publish** in the Command window.
- The dialog box on the next page will appear.
- The **Publish** dialog box:

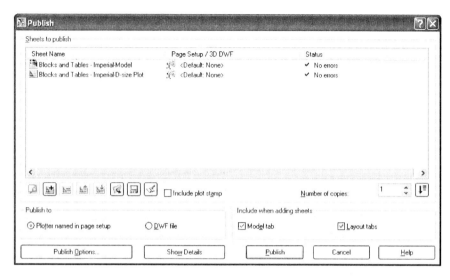

FIGURE 10-28

- As you can see, under **Sheets to publish** you will see a list of the current file's Model Space and layouts.
- *Select one of the sheets* and use the following buttons:

  - To preview the selected sheet (only single sheet) as we did in **Print Preview**.

  - To add more sheets from other drawings. The **Select Drawings** dialog box will be shown to select the desired file.

  - To remove one sheet or more from the list.

  - To move the sheet up in the list.

  - To move the sheet down in the list.

  - To open a sheet list saved previously.

- To save a sheet list for future printing.

- To set the Plot Stamp settings.
  - To include Plot Stamp in each sheet, click the **Include plot stamp** checkbox on.
  - Specify the number of copies.
  - Select to publish either to **Plotter named in page setup** or **DWF file** (select this option if you want to get a .dwf file).
  - When you add new sheets would you like to include the Model Space as one of the sheets or not?

  - Click the **Publish Options** button and you will see the following dialog box:

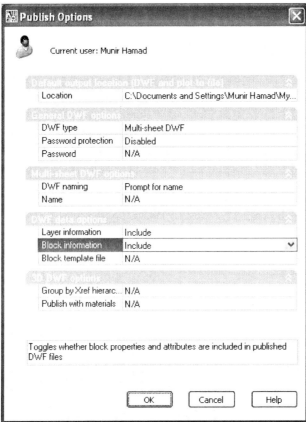

FIGURE 10-29

- Control the following settings:
  - Specify the location of the file.
  - Specify if the DWF file will be single sheet or multiple sheet.
  - Select to include a Password, or not, for the DWF file.
  - If you select multisheet, select either to prompt for the DWF naming or not.
  - Select whether to include the Layer and Block information (like attributes) in the DWF file.
  - If this is a 3D drawing, specify to **Group by Xref hierarchy** or not. And to **Publish with materials** or not.
- Once you are done, click **OK**.
- For the selected sheet you can see more details about it such as, Plot device, Plot size, Plot scale, etc.
- Once you are done setting all of these things, click the **Publish** button, which will lead to AutoCAD creating the sheets one by one.
- On the next page you can see an example of a multisheet DWF file.
- In order to see this file, simply double-click it and Windows will launch the Autodesk DWF Viewer automatically.
- An example of a multisheet DWF file:

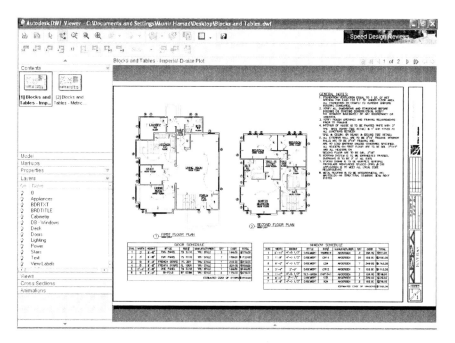

FIGURE 10-30

CREATING MULTIPLE-SHEET DWF FILES (METRIC & IMPERIAL)

 Workshop 21

1. Start AutoCAD 2008, and open the file: **Small_Villa_Ground_Floor_Plan_Metric_Workshop_21.dwg**. (If you solved the previous workshop correctly, open your file.) Or, open the file: **Small_Villa_Ground_Floor_Plan_Imperial_Workshop_21.dwg**. (If you solved the previous workshop correctly, open your file.)
2. Start the **Publish** command.
3. Select the Model Space sheet and remove it.
4. Select to **Add** new **sheets** and browse for **AutoCAD 2008\Sample\Blocks and Tables – Metric.dwg**.
5. Also, remove the Model space sheet.
6. Make the ISO A1 sheet the top sheet.
7. Select to Publish to DWF file.
8. Click **Publish** to create the file. It will ask you for the filename and place to save the file in, specify accordingly.
9. **Publish** will be done in the background. You can see this from the tray at the lower right corner of the AutoCAD 2008 window.
10. After the background publishing, browse to the place you saved the file in and doube-click it. Autodesk DWF Viewer will start automatically and show you the contents of the DWF file.

CHAPTER REVIEW

1. You should control what in Page Setup?
    a. Paper size
    b. Plotter to send to
    c. Viewports
    d. A & B
2. Layouts contain _____, which you can set for each its own scale.
3. DWF can be single sheet or multiple sheets.
    a. True
    b. False
4. You can choose to include or not to include the Layers in a DWF file.
    a. True
    b. False

5. Named Plot Style table file extension is:
   a. filename.ctb
   b. filename.stb
   c. filename.sbt
   d. filename.bct
6. In Name Plot Style table, if you assign a plot syle to a layer, you need to switch _____ from the Status bar to see this lineweight in the layout.

CHAPTER REVIEW ANSWERS

1. d
2. Viewports
3. a
4. a
5. b
6. LWT

APPENDIX A

ABOUT THE DVD

- Included on the DVD are exercise/project files from the text, figures from the text (many in color), third party software, and other files related to topics for AutoCAD 2008
- **Includes AutoCAD 2008 trial demo (30 day trial from date of installation)**
- See the "README" files for any specific information/system requirements related to each file folder, but most files will run on Windows XP or higher

APPENDIX B

HOW TO CREATE A TEMPLATE FILE

In This Chapter

- What are template files?
- What are the elements included in a template file?
- How to create a template file?

INTRODUCTION

- Companies using AutoCAD are always looking for two main things:
  - Unify their work to a certain standard (homemade or international).
  - Speed up the process of producing any drawing.
- The answer to these two issues is to create template files.
- Template files will ensure the decision makers in any company that all the premade settings for the drawings are already done in the templates, hence this will cut production time by minimum of 30%.
- Also, templates that are be handed to all users in any company will guarantee that these people will be using the same source. Accordingly, no personal initiative will be allowed, which will allow one system to prevail.
- Template files are °.dwt.

WHAT ARE THE ELEMENTS TO INCLUDE IN A TEMPLATE FILE?

- These are the elements to be included in a template file:
  - Drawing units
  - Drawing limits
  - Grid and Snap settings
  - OSNAP settings
  - POLAR settings
  - Layers
  - Linetypes

- Text Styles
- Dimension Styles
- Table Styles
- Layouts (including Border blocks and Viewports)
- Page Setups
- Plot Style tables
- **NOTE** No need to include blocks in the template file, but rather store them in files (each category in a separate file like Architectural, Civil, Mechanical, etc.).
- You can't save Tool Palettes inside a template file as Tool Palettes will be available for all files on a single computer.

HOW TO CREATE A TEMPLATE FILE?

- Start AutoCAD.
- Do the paperwork to prepare the above-mentioned settings. This step may need consultation with other people who operate AutoCAD in the same company.
- Create a new file using the simplest template file, **acad.dwt**.
- The new file will contain the minimum drawing file.
- Build inside this file all the above-mentioned elements.
- Once you are done, select from menus **File/Save As**. At **Files of type** select **AutoCAD Drawing Template (*.dwt)**, so AutoCAD will go directly to the **Template** folder under the AutoCAD folder.
- You can save this file in this folder, or you can create your own folder that will accommodate all of your template files.
- **NOTE** It is highly recommended to store your files in a different folder away from AutoCAD folders.
- You can create as many templates as you wish.
- If you want to edit an existing template, simply do the following:
  - Using menus select **File/Open** and at **Files of type** select **Drawing Template (*.dwt)**.
  - Open the desired template, do the changes.
  - Save it under the same name or use a new name.

APPENDIX C

INQUIRY COMMANDS

In This Chapter

- Why we need Inquiry commands
- ID command
- DIST command
- AREA command
- LIST command

INTRODUCTION

- These commands are to:
  - Identify a point coordinate
  - Measure a distance between two points
  - Calculate the area between points or of an object
  - List information about an object
- All of these commands are accessible from the **Inquiry** toolbar.

ID COMMAND

- To locate the coordinate of a point.
- Use one of the following methods to issue the command:
  - From the **Inquiry** command, click the **Locate Point** button.
  - From menus select **Tools/Inquiry/ID Point.**
  - Type **id** in the Command window.
- Either way AutoCAD will display the following prompt:

```
Specify point:
```

- Click on the desired point. AutoCAD will display something like:

```
X = 21.0000     Y = 11.5000     Z = 0.0000
```

DIST COMMAND

- To measure the distance between two points.
- Use one of the following methods to issue the command:
  - From the **Inquiry** command, click the **Distance** button.
  - From menus select **Tools/Inquiry/Distance.**
  - Type **dist** in the Command window.
- Either way AutoCAD will display the following prompt:

```
Specify first point:
Specify second point:
```

- Click on the two points desired. AutoCAD will display something like:

```
Distance = 10.0000, Angle in XY Plane = 0, Angle from XY Plane = 0
Delta X = 10.0000, Delta Y = 0.0000, Delta Z = 0.0000
```

AREA COMMAND

- To calculate the area between points or of an object.
- You can calculate area for:
  - Points, assuming there are lines connecting them.
  - Objects, like a circle or polyline.
- You can calculate two types of areas:
- *Simple* area (single area).
- *Complex* area (areas inside areas, and you want the net area).
- If you start the Area command and specify the points or select the object, it will calculate Simple area.
- To calculate Complex area you have to start with either **Add** or **Subtract**.
- Use one of the following methods to issue the command:
  - From the **Inquiry** command, click the **Area** button.
  - From menus select **Tools/Inquiry/Area.**
  - Type **area** in the Command window.
- Either way AutoCAD will display the following prompt:

```
Specify first corner point or [Object/Add/Subtract]:
```

Specify First Corner
- To calculate simple area consists of points assuming lines are connecting them. Specify the first point and AutoCAD will prompt:

  ```
  Specify next corner point or press ENTER for total:
  ```

- Keep on doing the same, up until you press [Enter]. The following message will appear:

  ```
  Area = 33.3750, Perimeter = 23.6264
  ```

Object
- To calculate the area by selecting an object like circle, or polyline, etc.,. Type **O** or right-click and select **Object**. AutoCAD will prompt:

  ```
  Select objects:
  ```

- Once you select the desired object, AutoCAD will report (the example is a circle):

  ```
  Area = 28.2743, Circumference = 18.8496
  ```

Add/Subtract
- You need these two modes in order to calculate Complex area, which is areas inside area, and you want the net area.
- Start with either one of these two modes and AutoCAD will assume that you are starting with Area = 0.00. Hence, you will add the outer area, then you can subtract the inner areas, or you can subtract the inner areas, then add the outer area.
- Assume we started with the **Add** mode. AutoCAD will prompt you:

  ```
  Specify first corner point or [Object/Subtract]:
  ```

- You can specify area(s) using either points or object. Whenever you are done, switch the mode to **Subtract** mode, and so on.
- While you are adding and subtracting AutoCAD will give you the current value of the area up until the last area added/subtracted.
- Once you are done, press [Enter] twice and AutoCAD will report to you the final value of the area.

LIST COMMAND

- To list information about a selected object without the ability to edit them.
- Use one of the following methods to issue the command:
  - From the **Inquiry** command, click the **List** button.

- From menus select **Tools/Inquiry/List.**
- Type **list** the Command window.
- Either way AutoCAD will display the following prompt:

```
Select objects:
```

- Select the desired object and AutoCAD will list the following information (the example is circle):

```
CIRCLE      Layer: "0"
            Space: Model space
             Handle = 8d
 center point, X = 30.5000  Y = 13.5000  Z = 0.0000
         radius    3.0000
    circumference  18.8496
            area   28.2743
```

NOTE
- Properties command is much better, as you will have the ability to edit any data you would like to change.

INDEX

2D Draw panel, 27, 31, 37, 79

A
Adaptive, 284
Adcenter, 137
Add plot style table button, 285
Add style button, 285
Add-a-plot style table wizard, 285
Adding viewports, 268–273
Alignment, 200
Alternate units, 224
Alternate units tab, 226
   After primary value, 226
   Below primary value, 226
   Display alternate units, 226
   Multiplier, 226
   Prefix, 226
   Round distance, 226
   Suffix, 226
Angular dimensions, 226, 242
Annotative, 164
ANSI hatches, 159
Arc command, 26, 31
Arcs, 38
Arc length, 242
Architectural tick, 215
Area command, 304
Array command, 110–111
   Angle of array, 111
   Number of columns, 111
   Rectangular, 111
   Select objects button, 111
Arrowhead tick, 215
Arrows tab, 215
Associative hatching and hatch origin (imperial), 168
Associative hatching and hatch origin (metric), 167–168
Associative hatching, 164, 167
Autodesk DWF viewer, 290
Autodesk plot style manager, 280
Add-a-plot style table wizard, 280

B
Base point, 103, 122, 131
Baseline dimensions, 244
Baseline spacing, 215
   Basic, 228
Bhatch command, 157–162
   Advanced features, 169–170
   Draw/Hatch, 157
   Hatch origin, 166
   Options, 164–165
   Preview hatch, 162
   Selecting hatch pattern, 157–160
Blocks, 129–141
   Base point, 131
   Block unit, 131
   Creating blocks, 129–133
   Inserting blocks, 133–136
   Pick point, 131
Block Automatic Scaling, 140–142
   Block unit, 141
   Format/Units, 140
   Inserted contents, 141
Border properties, 200
Borders tab,
   Color, 202
   Linetype, 202
   Lineweight, 202
Boundary retention, 169
Boundary set, 169–170

Break button, 119
Break command, 119–120
  Break button, 119
Break size, 218
Breaking objects, 120
BYLAYER, 57

C

C (Crossing), 100
Cartesian coordinates, 2
Center mark button, 247
Chamfer Command, 81–83
  Angle, 82
  Distances, 82
  Length and angle, 82
  Method, 83
  Multiple, 83
  Trim, 82
  Two distances, 81
Chamfering objects, 83
Change an object's properties, 65–67
Circle command, 29–31
  2 Points, 29
  3 Points, 29
  Center/Diameter, 29
  Center/Radius, 29
Clipping an existing viewport, 272–273
Color, 283
Color-dependent plot style table, 280–284
Continuous and baseline dimensions, 244
Continuous dimensions, 244
Controlling dimension styles, 230
Copy command, 104–105
  2D Draw, 104
  Modify/Copy, 105
Copying Objects, 105–106
CounterClockWise (CCW), 3
CP (Crossing Polygon), 101
Creating a block (imperial), 132–133
Creating a block (metric), 131–132
Creating a child style, 229–230

Creating a new file, 9
Creating a table style (imperial), 203
Creating a table style (metric), 202–203
Creating a text style (imperial), 185
Creating a text style (metric), 185
Creating dimension styles (imperial), 233
Creating dimension styles (metric), 231–232
Creating layouts and page setup (imperial), 266
Creating layouts and page setup (metric), 266
Creating multiple-sheet DWF files (metric & imperial), 296
Creating tool palettes from scratch, 145–146
Creating tool palettes from scratch, 145–146
Creating tool palettes using Design Center, 147
Crossing mode, 101
Crossing, 85
Ctb file, 284
Customizing tool palettes, 147
  How to copy a tool, 147
  How to customize a tool, 148–150
  How to paste a tool, 148

D

Dashboard, 4
Default to boundary extents, 167
Delete layer button, 61
Delta, 89
Deselect, 101
Design Center, 137–143
Deviation, 228
Diameter, 242
  Dim style, 249
Dim style, 249
  Dim text position, 249
Dim text position, 249
Dimangular, 241

Dimcenter, 247
Dimcontinue, 243
Dimension block properties, 249–250
Dimension blocks and grips, 248–249
Dimension break, 218
Dimension toolbar, 230
Dimension/leader, 246
Dimension style, 213–214
   Controlling dimension styles, 230
   Creating a child style, 229–230
   Creating dimension styles (imperial), 233
   Creating dimension styles (metric), 231–232
   Dimension toolbar, 213
   Dimstyle command, 213
   Format/dimension style, 213
   Lines tab, 214–217
   Linetype, 214
   Lineweight, 214
   Standard, 213
Dimension types, 210–212
   Angular, 211
   Art length, radius, and diameter, 211
   Baseline, 212
   Continuous, 211
   Linear and aligned, 210
   Ordinate, 212
   Quick leader, 212
Dimensioning commands, 234–235
   Aligned, 235
   Angular, 241
   Arc length, 236–238
   Baseline, 244
   Center mark, 247–248
   Continue, 243
   Diameter, 240–241
   Jogged, 240
   Linear, 234
   Ordinate, 238–239
   Quick dimension, 245
   Quick leader, 246
   Radius, 239

Dimensioning your drawing, 209–253
Dimlinear, 234
Dimradius, 239
Dimstyle command, 213
Direct Distance Entry, 24
Direction button, 55
Display viewport dialog, 269
Dist command, 304
Dither, 283
Drag-and-Drop, 139
Draw/Arc menu, 27
Draw/Circle menu, 30
Drawing a circle, 31–32
Drawing arcs, 28–29
Drawing limits, 5
Drawing limits, 55–56
Drawing Lines, 21–27
Drawing polylines, 39
Drawing the plan (imperial), 94–97
Drawing the plan (metric), 91–94
Drawing units, 53–55
   Angle, 54
   Clockwise, 54
   Format/Units, 54
   Length, 54
   Precision, 54
   Type, 54
Drawing commands, 73–76
   Modify/Offset, 74
   Offset button, 74
   Offset distance, 74
   Offset, 74
Drawing using OSNAP and OTRACK, 43
Drawing using OSNAP, 35–36
Drawing using polar tracking, 46–47
D-sized layout, 278
DTEXT command, 186–187
   Single-line text button, 186
Dwf files, 290
DYN button, 4
DYN, 20–21, 24–25
Dynamic drafting, 20–21
Dynamic, 90

E

Edit hatching (imperial), 178
Edit hatching (metric), 178
Edit/Undo, 49
Editing blocks, 151–154
    Add pick points, 161
    Add select objects, 161
    Add to working set, 152
    Refedit toolbar, 152
    Remove boundaries, 162
    Remove from working set, 152
    Selecting an area to be hatched, 161
    View selections, 162
Editing existing hatch, 176–178
    Edit hatch button, 176
    Hatchedit command, 176
    Recreate boundary, 177
Editing text (metric and imperial), 197–198
Equal X and Y spacing, 22
Erase command, 47–49
    Crossing, 48
    Modify, 47
    Modify/Erase, 47
    Window, 48
Example hatch, 164
Existing set, 170
Exploding blocks, 136–137
    2D draw, 136
    Explode button, 136
    Extend command, 86–89
    Extending objects, 89
    Extension lines, 215

F

F (Fence), 101
Fence, 85
Fill color, 200
Fill style, 284
Fillet command, 77–79
    Modify/Fillet, 79
    No trim, 78
    Trim, 78
Filleting objects, 80

Filter tree, 63
Find text string, 197
Find and replace, 196
    Edit/find, 196
    Find text string, 197
    Find, 196
    Text, 196
Fine tuning, 223
    Flip arrow, 249
Fit tab, 222
    Suppress arrows, 223
Flip arrow, 249
Format/Units, 134
Fraction height scale, 220
Freeze viewport, 276
Freezing layers in viewport, 275
Freezing layers in viewport, 275
Furniture layout, 143
Furniture, 251
Furniture, 277, 278

G

Gap tolerance, 170
General tab, 200
    Horizontal distances, 201
    Margins, 201
    Vertical distances, 201
Gradient command, 172–176
    Gradient button, 172
    Select color, 174–175
    Shade, 173
    Tint, 173
Grayscale, 283
Grid behavior, 22
Grid button, 22–23
Grid X spacing, 22
Grid Y spacing, 22
Grips, 121–126
    Mirror, 122
    Move, 122
    Rotate, 122
    Scale, 122
    Stretch, 122
Grips and DYN, 123
    Half-width, 38

H

Hatches,
    Create separate hatches, 165
    Draw order, 165
    Editing exiting hatch, 176–178
    Inherit properties, 166
Hatch origin, 166–167
Hatch patterns, 158–160
    Predefined, 159–160
    User defined, 158–159
Hatching, 157–180
Hatching using bhatch command (imperial), 163–164
Hatching using bhatch command (metric), 162–163
Hatching using tool palettes, 171–172
How to copy a tool, 147
How to create a new layout, 257–263
How to create a template file, 301–302
How to customize a tool, 148–150
How to paste a tool, 148

I

ID command, 303
    Area command, 304
    Dist command, 304
    List command, 305
Import text, 190
Inherit options, 170
Inherit properties, 166
In-place text editor, 188
Insert block button, 133
Insert block, 139
Insert/Block menu, 134
Inserting blocks (imperial), 136
Inserting blocks, 133–136
Inserting blocks (metric), 135
Inserting tables (imperial), 206
Inserting tables (metric), 205
Inserting viewport (imperial), 277–279
Inserting viewport (metric), 276–277
Insertion scale, 55
Inside, 252
Islands, 169
    Ignore, 169
    Islands detection, 169
    Normal, 169
    Outer, 169
ISO standard, 221
Isometric Snap, 23

J

Jog angle, 219
Jog height factor, 219
Jogged, 242

L

L (Last), 101
Layers, 56–65
    Format/Layer, 57
    How to add more layers, 61
    How to change object's layer, 63
    How to create a new layer, 57
    How to delete layers, 61
    How to make a layer the current layer, 60–61
    How to make an object's layer current, 63
    How to rename a layer, 62
    How to select layers, 61
    How to set a color for a layer, 58
    How to set a linetype for a layer, 59
    How to set a lineweight for a layer, 60
    New layer button, 57
Layers toolbar, 56–65
Layers panel, 63
    Make object's layer current button, 63
Layers properties manager, 60–66
Layer switches, 64
    On/off, 64
    Plot/no plot, 64
    Thaw/freeze, 64
    Unlock/lock, 64
Layer previous button, 65
Layer properties manager, 60–65
Layers properties manager, 276
Layers toolbar, 276

Layouts, 225, 255–263, 267
    How to create a new layout, 257–263
    Objects, 256
    Page setup, 256–257
    Page setup manager, 256
    Viewports, 256
Layout elements tab, 257
Layout wizard, 259–263
Leading, 225
    Length, 38
Lengthen command, 89–90
    Delta, 89
    Dynamic, 90
    Modify/Lengthen, 89
    Percent, 90
    Total, 90
Lengthening objects, 90–91
    Limits, 228
Limits, 5
Line command, 19–20
Line end style, 284
Line join style, 284
Linear and aligned dimensions, 236
Linear button, 234
Linear dimensions, 225
    Fraction format, 225
    Precision, 225
    Unit format, 225
Linear jog dimension, 219
Lines and arcs, 27
Lines tab, 214–217
Linetype, 59, 202, 284
Lineweight, 60, 202, 284
List command, 305
Load button, 59
Lwt button, 284
LWT button, 60

M

Major line spacing, 22
Match properties, 57
Maximize viewport button, 275
Measurement scale, 225
    Scale factor, 225

Mid, 32
Midpoint, 40
Mirror command, 116–117
Mirroring objects, 117
Model space mode, 273–274
Model space, 6, 255–256
Modifying commands, 99
Modifying viewports, 274–275
    Maximizing, 275
    Pan, 275
    Scaling, 274
    Zoom, 275
Move command, 102
    2D Draw, 102
    Modify/Move, 103
    Move button, 102
Moving Objects, 103–104
MTEXT, 234
MTEXT command, 187,
    Multiline text button, 187
    Multiple option, 75–76, 83
    Multiple rectangular viewport, 269
    Multiple, 79
    Multiple-page DWFs, 292
    Multiplier, 226

O

Object snap toolbar, 34
Object track (OTRACK), 40–43
Objects, 256
Oblique, 215
Offsetting objects, 76–77
ON/OFF, 56
One-point OTRACK, 41
Oops command, 49
Open an existing file, 10
ORTHO button, 25
Ortho tool, 24
OSNAP button, 34–35
OSNAP, 40
OSNAP settings, 36
    Center, 36
    Endpoint, 36

Midpoint, 36
Quadrant, 36
OTRACK, 40–43
Outside, 252

P
P (Previous), 101
Page setup manager, 256, 263–265, 284
Page setup, 256
Page size, 265
Pan realtime, 8
Paper button, 274
Paper mode, 277, 278
Paper orientation, 265
Paper space mode, 273–274
Paper space, 6, 255–256
Pen #, 283
Pen assignment, 284
Percent, 90
Pick points button, 168
Pline command, 36–39, 271
Plot command, 289
Plot offset, 265
Plot stamp, 294
Plot style table editor button, 285
Plot style tables, 279–290
    Add-a-plot style table wizard, 280
    Autodesk plot style manager, 280
Plot style table editor, 282–284
    Adaptive, 284
    Color, 283
    Dither, 283
    Fill style, 284
    Grayscale, 283
    Line end style, 284
    Line join style, 284
    Linetype, 284
    Lineweight, 284
    Page setup manager, 284
    Pen #, 283
    Pen assignment, 284
    Screening, 283
Inquiry commands, 303–307
    Area command, 304
    Dist command, 304
    ID command, 303
    List command, 305
Plot style table, 265
Plot style, 265
Plotter name, 294
Plotting your drawing, 255–296
Polar array, 115
Polar button, 44
Polar tracking, 43–47
    Additional angles, 44
    Increment angle, 44
    Polar angle settings, 44
Polar snap, 22, 45
Polygonal option, 272
Polygonal viewport button, 271
Polyline, 36–40
Precision, 249
Precision Method 3: Object Snap (OSNAP), 32–35
Precision Method, 22–25
Prefix, 226
Primary units, 224
Primary units tab, 220, 224
Print command, 256
Properties manager, 57
    Properties palette, 65–67
Publish button, 292–293
Publish options, 294–296
Putting dimensions on the plan (imperial), 251
Putting dimensions on the plan (metric), 250

Q
Qleader, 246
QNew button, 9
Quick dimension button, 245
Quick leader button, 246

R
Radius dimension jog, 219
Radius, 79, 242

Rectangular array, 112–115
   Center point, 114
   Methods and values, 114
   Polar, 113
   Rotate items as copied, 115
Rectangular snap, 23
Redo command, 49
Redraw and regen commands, 50–51
   Redraw, 50
   View/Redraw, 50
   View/Regen, 50
Rename option, 257
Right-clicking, 6
Rotate command, 106–107
Rotating objects, 107–108
Round distance, 226
Running OSNAP, 34–35

S
Saving files, 13
Scale Command, 108–109
Scale factor, 225
Scale lineweights, 265
Scaling Objects, 109–110
Scaling viewport (imperial), 277–279
Scaling viewport (metric), 276–277
Screening, 283
Search entire drawing, 197
Select drawings, 293
Selecting objects, 100–102
   All, 101
   C (Crossing), 100
   CP (Crossing Polygon), 101
   Crossing mode, 101
   Deselect, 101
   F (Fence), 101
   L (Last), 101
   Noun/Verb Selection, 101
   P (Previous), 101
   W (Window), 100
   Window Mode, 101
   WP (Window Polygon), 100
Set indents, 191
Settings, 22

Settings, 35
Sharing data between AutoCAD files using Design Center (blocks), 137–140
   Design Center button, 137
Sheets to publish, 293
Single-line text, 193
   Mutliline text, 194
Single polygonal viewport, 271
Single rectangular viewport, 269
Single-page DWFs, 291
Snap and Grid, 23–24
SNAP button, 22
Snap Y spacing, 22
Specified origin button, 167
Spelling check, 196
Standard annotation toolbar, 49, 289, 292
Stretch command, 117–118
Stretching objects, 119
Suffix, 226
Swatch, 159
   Symmetrical, 228
Symbols and arrows tab, 215–217
   Arc length symbol, 218
   Arrowheads, 218
   Break size, 218
   Center marks, 218
   Dimension break, 218

T
Table command, 203–205
   Draw/Table, 203
   Specify insertion point, 204
   Specify window, 205
   Table style name, 204
   Tables panel, 203
Table direction, 200
Table panel, 198
Table style, 198–202
   Alignment, 200
   Border properties, 200
   Data, 200
   Fill color, 200

    General tab, 200
    Header, 200
    Standard, 199
    Start with style, 199
    Starting table, 200
    Table direction, 200
    Table panel, 198
    Table style, 198
    Title, 200
Templates, 301–302
Text alignment, 221
Text angle, 202
Text and grips, 195
    MTEXT edit, 195
Text and tables, 181–207
Text appearance, 220
    Primary units tab, 220
    Text alignment, 221
    Text color, 220
    Text height, 220
    Text placement, 220
    Text style, 220
Text color, 202
Text formatting toolbar, 188
Text height, 202
Text placement, 220
Text properties, 201
Text style button, 182
Text style, 182–186, 202
    Annotative height, 183
    Effects, 184
    Font name, 183
    Font style, 183
    Standard, 182
    Text panel, 182
    Text style button, 182
Text tab, 201, 219
    Text angle, 202
    Text color, 202
    Text height, 202
    Text properties, 201
    Text style, 202
Thaw viewport, 276
Through point option, 75

Tolerance tab, 227
Tool palettes, 143–150
    Creating tool palettes from scratch, 145–146
    Creating tool palettes using Design Center, 147
    Using and customizing tool palettes (metric & imperial), 150–151
Tool/Palettes properties, 67
Tools/Spelling button, 196
Total, 90
Trim, 79
Trim command, 84–86
    2D draw, 84
    Crossing, 85
    Fence, 85
    Modify/Trim, 84
    Trim button, 84
Trimming objects, 86
Two-point OTRACK, 40

U

Undo Command, 49, 79
Undo option, 75
Uniform scale, 134
Unit format, 225
    Fraction format, 225
    Precision, 225
Using a template, 257
Using and customizing tool palettes (metric & imperial), 150–151
Using Design Center (imperial), 143
Using Design Center (metric), 142
Using grips, 125
Using layout command, 259

V

Viewports, 267–273
    Clipping an existing viewport, 272–273
    Converting an object to viewport, 272
    Modifying viewports, 274–275
    Multiple rectangular viewport, 269

Viewports, (cont.)
   Single polygonal viewport, 271
   Single rectangular viewport, 269
Vectors, 24
View/Pan, 8
View/Zoom, 8
Viewing commands, 8
Viewport spacing, 270, 278
Viewports toolbar, 269–270
Viewports, 256
Vports, 269

W

W (Window), 100
Width, 39
Window mode, 101
WP (Window Polygon), 100
Writing text (metric and imperial), 191

X

Xdatum option, 239

Z

Zero suppression, 225
Zoom all, 9
Zoom center, 8
Zoom dynamic, 8
Zoom extents, 9
Zoom in, 8
Zoom object, 8
Zoom out, 8
Zoom previous, 8
Zoom realtime, 8
Zoom scale, 8
Zoom toolbar, 8
Zoom window, 8
Zoom/Scale, 8